Universitext

Bernt Øksendal

Stochastic Differential Equations

An Introduction with Applications

Second Edition

Springer-Verlag Berlin Heidelberg New York
London Paris Tokyo Hong Kong

Bernt Øksendal
Department of Mathematics
University of Oslo
Box 1053, Blindern
N-0316 Oslo 3
Norway

The front cover shows the graph of a sample path of a geometric Brownian motion, i.e. the solution X_t of the (1-dimensional) stochastic differential equation

$$\frac{dX_t}{dt} = (r + \alpha \cdot W_t)X_t, \; X_0 = 1,$$

where W_t is white noise. This process is often used to model "exponential growth under uncertainty". See Chapters 5, 10 and 11.
The figure is a computer simulation for the case $\alpha = 0.4$, $r = 1.08$. The mean value of X_t, $\exp(rt)$, is also drawn, $0 \leq t \leq 3$. It was produced by courtesy of Tore Jonassen, University of Oslo.

AMS Subject Classification (1980): 60Hxx, 60G40, 60J45, 60J60, 93E11, 93E20

ISBN 3-540-51740-5 Springer-Verlag Berlin Heidelberg New York
ISBN 0-387-51740-5 Springer-Verlag New York Heidelberg Berlin

ISBN 3-540-15292-X 1. Auflage Springer-Verlag Berlin Heidelberg New York Tokyo
ISBN 0-387-15292-X 1st edition Springer-Verlag New York Heidelberg Berlin Tokyo

Printed in the United States of America

2141/3140-543210 – Printed on acid-free paper

To my family
Eva, Elise, Anders, and Karina

Preface

In the second edition I have split the chapter on diffusion processes in two, the new Chapters VII and VIII:
Chapter VII treats only those basic properties of diffusions that are needed for the applications in the last 3 chapters. The readers that are anxious to get to the applications as soon as possible can therefore jump directly from Chapter VII to Chapters IX, X and XI.

In Chapter VIII other important properties of diffusions are discussed. While not strictly necessary for the rest of the book, these properties are central in today's theory of stochastic analysis and crucial for many other applications.

Hopefully this change will make the book more flexible for the different purposes. I have also made an effort to improve the presentation at some points and I have corrected the misprints and errors that I knew about, hopefully without introducing new ones. I am grateful for the responses that I have received on the book and in particular I wish to thank Henrik Martens for his helpful comments.

Tove Lieberg has impressed me with her unique combination of typing accuracy and speed. I wish to thank her for her help and patience, together with Dina Haraldsson and Tone Rasmussen who sometimes assisted on the typing.

Oslo, August 1989 Bernt Øksendal

Preface to the First Edition

These notes are based on a postgraduate course I gave on stochastic differential equations at Edinburgh University in the spring 1982. No previous knowledge about the subject was assumed, but the presentation is based on some background in measure theory.

There are several reasons why one should learn more about stochastic differential equations: They have a wide range of applications outside mathematics, there are many fruitful connections to other mathematical disciplines and the subject has a rapidly developing life of its own as a fascinating research field with many interesting unanswered questions.

Unfortunately most of the literature about stochastic differential equations seems to place so much emphasis on rigor and completeness that is scares many nonexperts away. These notes are an attempt to approach the subject from the nonexpert point of view: Not knowing anything (except rumours, maybe) about a subject to start with, what would I like to know first of all? My answer would be:

1) In what situations does the subject arise?
2) What are its essential features?
3) What are the applications and the connections to other fields?

I would not be so interested in the proof of the most general case, but rather in an easier proof of a special case, which may give just as much of the basic idea in the argument. And I would be willing to believe some basic results without proof (at first stage, anyway) in order to have time for some more basic applications.

These notes reflect this point of view. Such an approach enables us to reach the highlights of the theory quicker and easier. Thus it is hoped that notes may contribute to fill a gap in the existing literature. The course is meant to be an appetizer. If it succeeds in awaking further interest, the reader will have a large selection of excellent literature available for the study of the whole story. Some of this literature is listed at the back.

In the introduction we state 6 problems where stochastic differential equations play an essential role in the solution. In Chapter II we introduce the basic mathematical notions needed for the mathematical model of some of these problems, leading to the concept of Ito integrals in Chapter III. In Chapter IV we develop the stochastic calculus (the Ito formula) and in Chapter V we use this to solve some stochastic differential equations, including the first two problems in the introduction. In Chapter VI we present a solution of the linear filtering problem (of which problem 3 is an example), using the stochastic calculus. Problem 4 is the Dirichlet problem. Although this is purely deterministic we outline in Chapters VII and VIII how the introduction of an associated Ito diffusion (i. e. solution of a stochastic differential equation) leads to a simple, intuitive and useful stochastic solution, which is the cornerstone of stochastic potential theory. Problem 5 is (a discrete version of) an optimal stopping problem. In Chapter IX we represent the state of a game at time t by an Ito diffusion and solve the corresponding optimal stopping problem. The solution involves potential theoretic notions, such as the generalized harmonic extension provided by the solution of the Dirichlet problem in Chapter VIII. Problem 6 is a stochastic version of F. P. Ramsey's classical control problem from 1928. In Chapter X we formulate the general stochastic control problem in terms of stochastic differential equations, and we apply the results of Chapters VII and VIII to show that the problem can be reduced to solving the (deterministic) Hamilton-Jacobi-Bellman equation. As an illustration we solve a problem about optimal portfolio selection.

After the course was first given in Edinburgh in 1982, revised and expanded versions were presented at Agder College, Kristiansand and University of Oslo. Every time about half of the audience have come from the applied section, the others being so-called "pure" mathematicians. This fruitful combination has created a broad variety of valuable comments, for which I am very grateful. I particularly wish to express my gratitude to K. K. Aase, L. Csink and A. M. Davie for many useful discussions.

I wish to thank the Science and Engineering Research Council, U. K. and Norges Almenvitenskapelige Forskningsråd (NAVF), Norway for their financial support. And I am greatly indebted to Ingrid Skram, Agder College and Inger Prestbakken, University of Oslo for their excellent typing – and their patience with the innumerable changes in the manuscript during these two years.

Oslo, June 1985 Bernt Øksendal

Note: Chapters VIII, IX, X of the First Edition have become Chapters IX, X, XI of the Second Edition

We have not succeeded in answering all our problems. The answers we have found only serve to raise a whole set of new questions. In some ways we feel we are as confused as ever, but we believe we are confused on a higher level and about more important things.

Posted outside the mathematics reading room,
Tromsø University

Contents

I. Introduction

To convince the reader that stochastic differential equations is an important subject let us mention some situations where such equations appear and can be used:

(A) Stochastic analogs of classical differential equations

If we allow for some randomness in some of the coefficients of a differential equation we often obtain a more realistic mathematical model of the situation.

PROBLEM 1. Consider the simple population growth model

$$\frac{dN}{dt} = a(t)N(t), \quad N(0) = A \tag{1.1}$$

where $N(t)$ is the size of the population at time t, and $a(t)$ is the relative rate of growth at time t. It might happen that $a(t)$ is not completely known, but subject to some random environmental effects, so that we have

$$a(t) = r(t) + \text{"noise"},$$

where we do not know the exact behaviour of the noise term, only its probability distribution. The function $r(t)$ is assumed to be non-random. How do we solve (1.1) in this case?

PROBLEM 2. The charge $Q(t)$ at time t at a fixed point in an electric circuit satisfies the differential equation

$$L \cdot Q''(t) + R \cdot Q'(t) + \frac{1}{C} \cdot Q(t) = F(t), \; Q(0) = Q_0, \; Q'(0) = I_0 \tag{1.2}$$

where L is inductance, R is resistance, C is capacitance and $F(t)$ the potential source at time t.

Again we may have a situation where some of the coefficients, say $F(t)$, are not deterministic but of the form

$$F(t) = G(t) + \text{"noise"}. \tag{1.3}$$

How do we solve (1.2) in this case?

More generally, the equation we obtain by allowing randomness in the coefficients of a differential equation is called a stochastic differential equation. This will be made more precise later. It is clear that any solution of a stochastic differential equation must involve some randomness, i.e. we can only hope to be able to say something about the probability distributions of the solutions.

(B) Filtering problems

PROBLEM 3. Suppose that we, in order to improve our knowledge about the solution, say of Problem 2, perform observations $Z(s)$ of $Q(s)$ at times $s \leq t$. However, due to inaccuracies in our measurements we do not really measure $Q(s)$ but a disturbed version of it:

(1.4) $\quad Z(s) = Q(s) + \text{"noise"}.$

So in this case there are two sources of noise, the second coming from the error of measurement.

The filtering problem is: What is the best estimate of $Q(t)$ satisfying (1.2), based on the observations (1.4), where $s \leq t$? Intuitively, the problem is to "filter" the noise away from the observations in an optimal way.

In 1960 Kalman and in 1961 Kalman and Bucy proved what is now known as the Kalman-Bucy filter. Basically the filter gives a procedure for estimating the state of a system which satisfies a "noisy" linear differential equation, based on a series of "noisy" observations.

Almost immediately the discovery found applications in aerospace engineering (Ranger, Mariner, Apollo etc.) and it now has a broad range of applications.

Thus the Kalman-Bucy filter is an example of a recent mathematical discovery which has already proved to be useful - it is not just "potentially" useful.

It is also a counterexample to the assertion that "applied mathematics is bad mathematics" and to the assertion that "the only really useful mathematics is the elementary mathematics". For the Kalman-Bucy filter - as the whole subject of stochastic differential equations - involves advanced, interesting and first class mathematics.

(C) Stochastic approach to deterministic boundary value problems

PROBLEM 4. The most celebrated example is the stochastic solution of the Dirichlet problem:

> Given a (reasonable) domain U in $\mathbb{R}^n$ and a continuous function f on the boundary of U, ∂U. Find a function $\tilde{f}$ continuous on the closure $\bar{U}$ of U such that
>
> (i) $\tilde{f} = f$ on ∂U
>
> (ii) $\tilde{f}$ is harmonic in U, i.e.
>
> $$\Delta \tilde{f} = \sum_{i=1}^{n} \frac{\partial^2 \tilde{f}}{\partial x_i^2} = 0 \quad \text{in} \quad U.$$

In 1944 Kakutani proved that the solution could be expressed in terms of Brownian motion (which will be constructed in Chapter II): $\tilde{f}(x)$ is the expected value of f at the first exit point from U of the Brownian motion starting at $x \in U$.

It turned out that this was just the tip of an iceberg: For a large class of semielliptic 2nd order partial differential equations the corresponding Dirichlet boundary value problem can be solved using a stochastic process which is a solution of an associated stochastic differential equation.

(D) Optimal stopping

PROBLEM 5. Suppose a person has an asset or resource (e.g. a house, stocks, oil...) that she is planning to sell. The price X_t at time t of her asset on the open market varies according to a stochastic differential equation of the same type as in Problem 1:

$$\frac{dX_t}{dt} = rX_t + \alpha X_t \cdot \text{"noise"}$$

where r, α are known constants. The inflation rate (discounting factor) is a known constant ρ. At what time should she decide to sell?

We assume that she knows the behaviour of X_s up to the present time t, but because of the noise in the system she can of course never be sure at the time of the sale that her choice of time will turn out to be the best. So what we are searching for is a stopping strategy that gives the best result in the long run, i.e. maximizes the expected profit when the inflation is taken into account.

This is an optimal stopping problem. It turns out that the solution can be expressed in terms of the solution of a corresponding boundary value problem (Problem 4), except that the boundary is unknown (free) as well and this is compensated by a double set of boundary conditions.

(E) Stochastic control

PROBLEM 6. A stochastic analog of the "How much should a nation save?"-problem of F.P. Ramsey from 1928 (see Ramsey [1]) in economics is the following:

The basic economic quantities are

$K(t)$ = capital at time t

$L(t)$ = labour at time t

$P(t)$ = production rate at time t

$C(t)$ = consumption rate at time t

$U(C)\Delta t$ = the "utility" obtained by consuming goods at the consumption rate C during the time interval Δt.

Let us assume that the relation between $K(t)$, $L(t)$ and $P(t)$ is of the Cobb-Douglas form:

$$P(t) = AK(t)^{\alpha}L(t)^{\beta}, \tag{1.5}$$

where A, α, β are constants.

Further, assume that

$$\frac{dK}{dt} = P(t) - C(t) \tag{1.6}$$

and

$$\frac{dL}{dt} = a(t)\cdot L(t), \tag{1.7}$$

where $a(t) = r(t) +$ "noise" is the rate of growth of the population (labour).

Given a utility function U and a "bequest" function ψ, the problem is to determine at each time t the size of the consumption rate $C(t)$ which maximizes the expected value of the total utility up to a future time $T \le \infty$:

$$\max\{E[\int_0^T U(C(t))e^{-\rho t}dt] + \psi(K(T))\} \tag{1.8}$$

where ρ is a discounting factor.

II. Some Mathematical Preliminaries

Having stated the problems we would like to solve, we now proceed to find reasonable mathematical notions corresponding to the quantities mentioned and mathematical models for the problems. In short, here is a first list of the notions that need a mathematical interpretation:

(1) A random quantity

(2) Independence

(3) Parametrized (discrete or continuous) families of random quantities

(4) What is meant by a "best" estimate in the filtering problem (Problem 3)?

(5) What is meant by an estimate "based on" some observations (Problem 3)?

(6) What is the mathematical interpretation of the "noise" terms?

(7) What is the mathematical interpretation of the stochastic differential equations?

In this chapter we will discuss (1) - (3) briefly. In the next chapter (III) we will consider (6), which leads to the notion of an Ito stochastic integral (7).

In Chapters IV, V we consider the solution of stochastic differential equations and then return to a solution of Problem 1. In Chapter VI we consider (4) and (5) and sketch the Kalman-Bucy solution to the linear filtering problem. In Chapters VII and VIII we investigate further the properties of a solution of a stochastic differential equation. Then in Chapters IX, X and XI this is applied to solve the generalized Dirichlet problem, optimal stopping problems and stochastic control problems, respectively.

The mathematical model for a random quantity is a random variable:

DEFINITION 2.1. A random variable is an $\mathcal{F}$-measurable function $X:\Omega\to\mathbb{R}^n$, where $(\Omega,\mathcal{F},P)$ a is (complete) probability space and $\mathbb{R}^n$ denotes n-dimensional Euclidean space. (Thus $\mathcal{F}$ is a σ-algebra of subsets of Ω, P is a probability measure in Ω, assigning values in $[0,1]$ to each member of $\mathcal{F}$ and if B is a Borel set in $\mathbb{R}^n$ then $X^{-1}(B)\in\mathcal{F}$.)

Every random variable induces a measure μ_X on $\mathbb{R}^n$, defined by

$$\mu_X(B) = P(X^{-1}(B)).$$

μ_X is called the distribution of X.

The mathematical model for independence is the following:

DEFINITION 2.2. Two subsets $A, B, \in \mathcal{F}$ are called independent if

$$P(A\cap B) = P(A)\cdot P(B).$$

A collection $\mathcal{A} = \{\mathcal{H}_i;\ i \in I\}$ of families $\mathcal{H}_i$ of measurable sets is independent if

$$P(A_{i_1} \cap \ldots \cap A_{i_k}) = P(A_{i_1}) \ldots P(A_{i_k})$$

for all choices of $A_{i_1} \in \mathcal{H}_{i_1}, \ldots, A_{i_k} \in \mathcal{H}_{i_k}$.

The σ-algebra $\mathcal{H}_X$ induced by a random variable X is

$$\mathcal{H}_X = \{X^{-1}(B); B \in \mathcal{B}\},$$

where $\mathcal{B}$ is the Borel σ-algebra on $\mathbb{R}^n$.

A collection of random variables $\{X_i\ ;\ i \in I\}$ is independent if the collection of induced σ-algebras $\mathcal{H}_{X_i}$ is independent.

DEFINITION 2.3. A stochastic process is a parametrized collection of random variables

$$\{X_t\}_{t\in T}$$

defined on a probability space $(\Omega,\mathcal{F},P)$ and assuming values in $\mathbb{R}^n$.
The parameter space T is usually (as in this book) the halfline $[0,\infty)$, but it may also be an interval $[a,b]$, the non-negative integers and even subsets of $\mathbb{R}^n$ for $n\geq 1$.
Note that for each $t \in T$ fixed we have a random variable

$$\omega \to X_t(\omega)\quad ;\quad \omega \in \Omega.$$

On the other hand, fixing $\omega \in \Omega$ we can consider the function

$$t \to X_t(\omega)\quad ;\quad t \in T$$

which is called a path of X_t.
It may be useful for the intuition to think of t as "time" and each ω as an individual "particle" or "experiment". With this picture $X_t(\omega)$ would represent the position (or result) at time t of the particle (experiment) ω. Sometimes it is convenient to write $X(t,\omega)$ instead of $X_t(\omega)$. Thus we may also regard the process as a function of two variables

$$(t,\omega) \to X(t,\omega)$$

from $T\times\Omega$ into $\mathbb{R}^n$. This is often a natural point of view in stochastic analysis, because (as we shall see) there it is crucial to have $X(t,\omega)$ jointly measurable in (t,ω).

Finally we note that we may identify each ω with the function $t \to X_t(\omega)$ from T into $\mathbb{R}^n$. Thus we may regard Ω as a subset of the space $\tilde{\Omega} = (\mathbb{R}^n)^T$ of all functions from T into $\mathbb{R}^n$. Then the σ-algebra $\mathcal{F}$ will contain the σ-algebra $\mathcal{B}$ generated by sets of the form

$$\{\omega;\omega(t_1) \in F_1,\ldots,\omega(t_k) \in F_k\} \quad , \quad F_i \subset \mathbb{R}^n \text{ Borel sets}$$

($\mathcal{B}$ is the same as the Borel σ-algebra on $\bar{\Omega}$ if $T = [0,\infty)$ and $\tilde{\Omega}$ is given the product topology). Therefore one may also adopt the point of view that a stochastic process is <u>a probability measure P on the measurable space $((\mathbb{R}^n)^T,\mathcal{B})$</u>.

The <u>finite-dimensional distributions</u> of the process $X = \{X_t\}_{t\in T}$ are the measures $\mu_{t_1,\ldots,t_k}$ defined on $\mathbb{R}^{nk}$ by

$$\mu_{t_1,\ldots,t_k}(F_1\times F_2\times\ldots\times F_k) = P[X_{t_1}\in F_1,\ldots,X_{t_k}\in F_k] \; ; \; t_i\in T .$$

Here $F_1,\ldots,F_k$ denote Borel sets in $\mathbb{R}^n$.

The family of all finite-dimensional distributions determine many (but not all) important properties of the process X.

Conversely, given a family $\{\nu_{t_1,\ldots,t_k}; \; k \in N, \; t_i \in T\}$ of probability measures om $\mathbb{R}^{nk}$ it is important to be able to construct a stochastic process $Y = \{Y_t\}_{t\in T}$ having $\nu_{t_1,\ldots,t_k}$ as its finite-dimensional distributions. One of Kolomogorov's famous theorems states that this can be done provided $\{\nu_{t_1,\ldots,t_k}\}$ satisfies two natural consistency conditions: (See Lamperti [1])

<u>THEOREM 2.4</u>. (Kolmogorov's extension theorem).
For all $t_1,\ldots,t_k \in T$, $k\in \mathbb{N}$ let $\nu_{t_1,\ldots,t_k}$ be probability measures on $\mathbb{R}^{n\times k}$ s.t.

(K1) $$\nu_{t_{\sigma(1)},\ldots,t_{\sigma(k)}}(F_1\times\ldots\times F_k) = \nu_{t_1,\ldots,t_k}(F_{\sigma^{-1}(1)}\times\ldots\times F_{\sigma^{-1}(k)})$$

for all permutations σ on $\{1,2,\ldots,k\}$ and

(K2) $\nu_{t_1,..,t_k}(F_1\times..\times F_k) = \nu_{t_1,..,t_k,t_{k+1},..,t_{k+m}}(F_1\times..F_k\times\mathbb{R}^n\times..\times\mathbb{R}^n)$ for all $m \in \mathbb{N}$, where (of course) the set on the right hand side has a total of $k+m$ factors.

Then there exists a probability space $(\Omega,\mathcal{F},P)$ and a stochastic process $\{X_t\}$ on Ω, $X_t: \Omega \to \mathbb{R}^n$, s.t. $\nu_{t_1,...,t_k}(F_1\times...\times F_k) = P[X_{t_1} \in F_1,...,X_{t_k} \in F_k]$, for all $t_i \in T$, $k \in \mathbb{N}$ and all Borel sets F_i.

An important example: Brownian motion

In 1828 the Scottish botanist Robert Brown observed that pollen grains suspended in liquid performed an irregular motion. The motion was later explained by the random collisions with the molecules of the liquid. To describe the motion mathematically it is natural to use the concept of a stochastic process $B_t(\omega)$, interpreted as the position at time t of the pollen grain ω. We will generalize slightly and consider an n-dimensional analog.

To construct $\{B_t\}_{t\geq 0}$ it suffices, by the Kolmogorov extension theorem, to specify a family $\{\nu_{t_1,...t_k}\}$ of probability measures satisfying (K1) and (K2). These measures will be chosen so that they agree with our observations of the pollen grain behaviour:

Fix $x \in \mathbb{R}^n$ and define

$$p(t,x,y) = (2\pi t)^{-n/2}\cdot \exp\left(-\frac{|x-y|^2}{2t}\right) \quad \text{for } y \in \mathbb{R}^n,\ t>0$$

If $0\leq t_1\leq t_2\leq ...\leq t_k$ define a measure $\nu_{t_1,...,t_k}$ on $\mathbb{R}^{nk}$ by

(2.1) $$\nu_{t_1,...,t_k}(F_1\times..\times F_k) =$$

$$\int_{F_1\times..\times F_k} p(t_1,x,x_1)p(t_2-t_1,x_1,x_2)\ ...p(t_k-t_{k-1},x_{k-1},x_k)dx_1...dx_k$$

where we use the notation $dy=dm(y)$ for Lebesgue measure and the convention that $p(0,x,y)m(y) = \delta_x(y)$, the unit point mass at x.

Extend this definition to all finite sequences of t_i's by using (K1). Since $\int_{R^n} p(t,x,y)dy = 1$ for all $t \geq 0$, (K2) holds, so by Kolmogorov's theorem there exists a probability space $(\Omega,\mathcal{F},P^x)$ and a stochastic process $\{B_t\}_{t\geq 0}$ on Ω such that the finite-dimensional distributions of B_t are given by (2.1), i.e.

$$(2.2) \qquad P^x(B_{t_1} \in F_1,\ldots,B_{t_k} \in F_k) =$$

$$\int_{F_1\times\ldots\times F_k} p(t_1,x,x_1)\ldots p(t_k-t_{k-1}, x_{k-1},x_k)\, dx_1\ldots dx_k.$$

Such a process is called (a version of) <u>Brownian motion starting at x</u> (observe that $P^x(B_0 = x) = 1$).

The Brownian motion thus defined is not unique, i.e. there exist several quadruples $(B_t,\Omega,\mathcal{F},P^x)$ such that (2.3) holds.

However, for our purposes this is not important, we may simply choose any version to work with. As we shall soon see, the paths of a Brownian motion are (or, more correctly, can be chosen to be) continuous, a.s. Therefore we may identify (a.a.) $\omega \in \Omega$ with a continuous function $t \to B_t(\omega)$ from $[0,\infty)$ into $\mathbb{R}^n$. Thus we may adopt the point of view that Brownian motion is just the space $C([0,\infty), \mathbb{R}^n)$ equipped with certain probability measures P^x (given by (2.2) and (2.3) above). This version is called the <u>canonical</u> Brownian motion. Besides having the advantage of being intuitive, this point of view is useful for the further analysis of measures on $C([0,\infty), \mathbb{R}^n)$, since this space is Polish (i.e. a complete separable metric space). See Stroock and Varadhan [1].

We state some basic properties of Brownian motion:

(i) B_t is a <u>Gaussian process</u>, i.e. for all $0\leq t_1\leq\ldots\leq t_k$ the random variable $Z = (B_{t_1},\ldots,B_{t_k}) \in \mathbb{R}^{nk}$ has a <u>(multi)normal distribution</u>. This means that there exists a vector $M \in \mathbb{R}^{nk}$ and a non-negative definite matrix $C = [c_{jk}] \in \mathbb{R}^{nk\times nk}$ (the set of all $nk \times nk$ - matrices with real entries) such that

$$(2.3) \qquad E^x[\exp(i \sum_{j=1}^{nk} u_j Z_j)] = \exp(-\tfrac{1}{2} \sum_{j,m} u_j c_{jm} u_m + i \sum_j u_j M_j)$$

for all $u=(u_1,\ldots,u_{nk}) \in \mathbb{R}^{nk}$, where $i = \sqrt{-1}$ is the imaginary unit and E^x denotes expectation with respect to P^x. Moreover, if (2.3) holds then

(2.4) $M = E^x[Z]$ is the mean value of Z

and

(2.5) $c_{jk} = E^x[(Z_j-M_j)(Z_k-M_k)]$ is the covariance matrix of Z. (See Appendix A).

To see that (2.3) holds for $Z = (B_{t_1},\dots,B_{t_k})$ we calculate its left hand side explicity by using (2.2) (see Appendix A) and obtain (2.3) with

(2.6) $M = E^x[Z] = (x,x,\dots,x) \in \mathbb{R}^{nk}$

and

(2.7)
$$C = \begin{pmatrix} t_1 I_n & t_1 I_n & \cdots & t_1 I_n \\ t_1 I_n & t_2 I_n & \cdots & t_2 I_n \\ \vdots & \vdots & & \vdots \\ t_1 I_n & t_2 I_n & \cdots & t_k I_n \end{pmatrix}$$

Hence

(2.8) $\underline{E^x[B_t] = x}$ for all $t \geq 0$

and

(2.9) $\underline{E^x[(B_t-x)^2] = nt}$, $\underline{E^x[(B_t-x)(B_s-x)] = n\min(s,t)}$.

Moreover,

(2.10) $E^x[(B_t-B_s)^2] = n(t-s)$ if $t \geq s$,

since

$$E^x[(B_t-B_s)^2] = E^x[(B_t-x)^2-2(B_t-x)(B_s-x)+(B_s-x)^2]$$
$$= n(t-2s+s) = n(t-s), \quad \text{when } t \geq s.$$

(ii) B_t has <u>independent increments</u>, i.e.

(2.11) $B_{t_1}, B_{t_2}-B_{t_1},\dots,B_{t_k}-B_{t_{k-1}}$ are independent for all $0 \leq t_1 < t_2 \dots < t_k$.

To prove this we use the fact that normal random variables are independent iff they are uncorrelated. (See Appendix A). So it is enough to prove that

(2.12) $$E^x[(B_{t_i}-B_{t_{i-1}})(B_{t_j}-B_{t_{j-1}})] = 0 \quad \text{when } t_i < t_j,$$

which follows from the form of A:

$$E^x[B_{t_i}B_{t_j}-B_{t_{i-1}}B_{t_j}-B_{t_i}B_{t_{j-1}}+B_{t_{i-1}}B_{t_{j-1}}]$$
$$= n(t_i - t_{i-1} - t_i + t_{i-1}) = 0.$$

(iii) Finally we ask: Is $t \to B_t(\omega)$ continuous for almost all ω? Stated like this the question does not make sense, because the set $H = \{\omega;\ t \to B_t(\omega)$ is continuous$\}$ is not measurable with respect to the Borel σ-algebra $\mathcal{B}$ on $(\mathbb{R}^n)^{[0,\infty)}$ mentioned above (H involves an uncountable number of t's). However, if modified slightly the question can be given a positive answer. To explain this we need the following important concept:

DEFINITION 2.5. Suppose that $\{X_t\}$ and $\{Y_t\}$ are stochastic processes on $(\Omega,\mathcal{F},P)$. Then we say that $\{X_t\}$ is a version of $\{Y_t\}$ if

$$P(\{\omega;\ X_t(\omega) = Y_t(\omega)\}) = 1 \quad \text{for all } t.$$

Note that if X_t is a version of Y_t, then X_t and Y_t have the same finite-dimensional distributions. Thus from the point of view that a stochastic process is a probability law on $(\mathbb{R}^n)^{[0,\infty)}$ two such processes are indistinguishable, but nevertheless their path properties may be different.

The continuity question of Brownian motion can be answered by using another famous theorem of Kolmogorov:

THEOREM 2.6. (Kolmogorov's continuity theorem). Suppose that the process $X = \{X_t\}_{t\geq 0}$ satisfies the following condition:
For all $T>0$ there exist positive constants α, β, D such that

(2.13) $$E[|X_t-X_s|^\alpha] \leq D\cdot|t-s|^{1+\beta}\ ;\ 0 \leq s,\ t \leq T.$$

Then there exists a path-continuous version of X.

(See for example Stroock and Varadhan [1], p. 51).

For Brownian motion B_t it is not hard to prove that

(2.14) $$E^x[|B_t - B_s|^4] = n(n+2)|t-s|^2.$$

So Brownian motion satisfies Kolomogorov's condition (2.13) with $\alpha = 4$, $D = n(n+2)$ and $\beta = 1$, and therefore it has a path continuous version. From now on we will assume that B_t is such a continuous version.

Finally we note that

(2.15) If $B_t = (B_t^{(1)},\ldots,B_t^{(n)})$ is n-dimensional Brownian motion, then the 1-dimensional processes $\{B_t^{(j)}\}_{t\geq 0}$, $1 \leq j \leq n$ are independent, 1-dimensional Brownian motions.

III. Ito Integrals

We now turn to the question of finding a reasonable mathematical interpretation of the "noise" term in the equation of Problem 1:

$$\frac{dN}{dt} = (r(t)+\text{"noise"})N(t)$$

or more generally in equations of the form

$$\frac{dX}{dt} = b(t,X_t) + \sigma(t,X_t) \cdot \text{"noise"}, \tag{3.1}$$

where b and σ are some given functions.
Let us first concentrate on the case when the noise is 1-dimensional. It is reasonable to look for some stochastic process W_t to represent the noise term, so that

$$\frac{dX}{dt} = b(t,X_t) + \sigma(t,X_t) \cdot W_t. \tag{3.2}$$

Based on many situations, for example in engineering, one is led to assume that W_t has, at least approximately, these properties:

(i) $t_1 \neq t_2 \Rightarrow W_{t_1}$ and W_{t_2} are independent

(ii) $\{W_t\}$ is stationary, i.e. the joint distribution of $\{W_{t_1+t},\ldots,W_{t_k+t}\}$ does not depend on t.

(iii) $E[W_t] = 0$ for all t.

However, it turns out there does not exist any "reasonable" stochastic process satisfying (i) and (ii): Such a W_t cannot have continuous paths. If we require $E[W_t^2] = 1$ then the function $(t,\omega) \to W_t(\omega)$ cannot even be measurable, with respect to the σ-algebra $\mathcal{B}\times\mathcal{F}$, where $\mathcal{B}$ is the Borel σ-algebra on $[0,\infty]$. (See Kallianpur [1], p. 10).

Nevertheless it is possible to represent W_t as a generalized stochastic process called the white noise process.

That the process is generalized means that it can be constructed as a probability measure on the space $\mathcal{S}'$ of tempered distributions on $[0,\infty)$, and not as a probability measure on the much smaller space $\mathbb{R}^{[0,\infty)}$, like an ordinary process can. See e.g. Hida [1], Rozanov [1].

We will avoid this kind of construction and rather try to rewrite equation (3.2) in a form that suggests a replacement of W_t by a proper stochastic process:

Let $0 = t_0 < t_1 < \ldots < t_m = T$ and consider a discrete version of (3.2):

$$X_k - X_{k-1} = b(t_k,X_k)\Delta t_k + \sigma(t_k,X_k)W_k\Delta t_k, \tag{3.3}$$

where $X_j = X(t_j),\ W_k = W_{t_k},\ \Delta t_k = t_k - t_{k-1}.$

We abandon the W_k-notation and replace $W_k\Delta t_k$ by $\Delta V_k = V_k - V_{k-1}$, where $\{V_t\}_{t>0}$ is some suitable stochastic process. The assumptions (i), (ii) and (iii) on W_t suggest that V_t should have stationary independent increments with mean 0. It turns out that the only such process with continuous paths is the Brownian motion B_t. (See Knight [1]). Thus we put $V_t = B_t$ and obtain from (3.3):

$$X_k = X_0 + \sum_{j=1}^{k} b(t_j,X_j)\Delta t_j + \sum_{j=1}^{k} \sigma(t_j,X_j)\Delta B_j. \tag{3.4}$$

Is it possible to prove that the limit of the right hand side of (3.4) exists, in some sense, when $\Delta t_j \to 0$? If so, then by applying the usual integration notation we would obtain

$$X_t = X_0 + \int_0^t b(s,X_s)ds + \int_0^t \sigma(s,X_s)dB_s \tag{3.5}$$

and we would adopt as a convention that (3.2) really means that $X_t = X_t(\omega)$ is a stochastic process satisfying (3.5).

Thus, in the remainder of this chapter we will prove the existence of

$$\int_0^t f(s,\omega)dB_s(\omega)$$

where $B_t(\omega)$ is 1-dimensional Brownian motion starting at the origin, for a wide class of functions $f: [0,\infty] \times \Omega \to \mathbb{R}$. Then, in Chapter V, we will return to the solution of (3.5).

Suppose $0 \le S < T$ and $f(t,\omega)$ is given. We want to define

$$\int_S^T f(t,\omega)dB_t(\omega). \tag{3.6}$$

It is reasonable to start with a definiton for a simple class of functions f and then extend by some approximation procedure. Thus, let us first assume that f has the form

(3.7) $$\phi(t,\omega) = \sum_{j\geq 0} e_j(\omega) \cdot \chi_{[j\cdot 2^{-n}, (j+1)2^{-n})}(t),$$

where χ denotes the characteristic (indicator) function. For such functions it is reasonable to define

(3.8) $$\int_S^T \phi(t,\omega)dB_t(\omega) = \sum_{j\geq 0} e_j(\omega)[B_{t_{j+1}} - B_{t_j}](\omega),$$

$$\text{where } t_k = t_k^{(n)} = \begin{cases} k\cdot 2^{-n} & \text{if } S \leq k\cdot 2^{-n} \leq T \\ S & \text{if } k\cdot 2^{-n} < S \\ T & \text{if } k\cdot 2^{-n} > T \end{cases}$$

However, without any further assumptions on the functions $e_j(\omega)$ this leads to difficulties, as the next example shows.

EXAMPLE 3.1.

Choose
$$\phi_1(t,\omega) = \sum_{j\geq 0} B_{j\cdot 2^{-n}}(\omega) \cdot \chi_{[j\cdot 2^{-n}, (j+1)2^{-n})}(t)$$
$$\phi_2(t,\omega) = \sum_{j\geq 0} B_{(j+1)2^{-n}}(\omega) \cdot \chi_{[j\cdot 2^{-n}, (j+1)2^{-n})}(t)$$

Then

$$E[\int_0^T \phi_1(t,\omega)dB_t(\omega)] = \sum_{j\geq 0} E[B_{t_j}(B_{t_{j+1}} - B_{t_j})] = 0,$$

since $\{B_t\}$ has independent increments. (Here-and in the following-E means the same as E^0).

But

$$E[\int_0^T \phi_2(t,\omega)dB_t(\omega)] = \sum_{j\geq 0} E[B_{t_{j+1}} \cdot (B_{t_{j+1}} - B_{t_j})]$$
$$= \sum_{j\geq 0} E[(B_{t_{j+1}} - B_{t_j})^2] = T, \quad \text{by (2.10)}.$$

So, in spite the fact that both ϕ_1 and ϕ_2 appear to be very reasonable approximations to

$$f(t,\omega) = B_t(\omega),$$

their integrals according to (3.8) are not close to each other at all, no matter how large n is chosen.

This only reflects the fact that the variations of the paths of B_t are too big to enable us to define the integral (3.6) in the Riemann-Stieltjes sense. In fact one can show that the paths $t \to B_t$ of Brownian motion are nowhere differentiable, almost surely (a.s.). (See Breiman [1]). In particular, the total variation of the path is infinite, a.s.

In general it is natural to approximate a given function $f(t,\omega)$ by

$$\sum_j f(t_j^*, \omega) \cdot \chi_{[t_j,t_{j+1})}(t)$$

where the points t_j^* belong to the intervals $[t_j,t_{j+1})$, and then define $\int_S^T f(t,\omega)dB_t(\omega)$ as the limit (in a sense that we will explain) of $\sum_j f(t_j^*,\omega)[B_{t_{j+1}} - B_{t_j}](\omega)$ as $n \to \infty$. However, the example above shows that - unlike the Riemann-Stieltjes integral - it does make a difference here what points t_j^* we choose. The following two choices have turned out to be the most useful ones:

1) $t_j^* = t_j$ (the left end point), which leads to the <u>Ito integral</u>

2) $t_j^* = \frac{1}{2}(t_j + t_{j+1})$ (the mid point), which leads to the <u>Stratonovich integral</u>

In the end of this chapter we will explain why these choices are the best and discuss the relations and distinctions between the corresponding integrals.

In any case one must restrict oneself to a special class of functions $f(t,\omega)$ in (3.6), also if they have the particular form (3.7), in order to obtain a reasonable definition of the integral.
We will here present Ito's choice $t_j^* = t_j$. The approximation procedure indicated above will work out successfully provided that f has the property that each of the functions $\omega \to f(t_j,\omega)$ <u>only depends on the behaviour of $B_s(\omega)$ up to time t_j</u>.
This leads to the following important concepts:

<u>DEFINITION 3.2</u> Let $B_t(\omega)$ be n-dimensional Brownian motion. Then we define $\mathcal{F}_t = \mathcal{F}_t^{(n)}$ to be the σ-algebra generated by the random variables $B_s(\cdot)$; $s \leq t$. In other words, $\mathcal{F}_t$ is the smallest σ-algebra containing all sets of the form

$$\{\omega;\ B_{t_1}(\omega) \in F_1, \ldots, B_{t_k}(\omega) \in F_k\},$$

where $k=1,2,\ldots$, $t_j \geq 0$ and $F_j \subset \mathbb{R}^n$ are Borel sets. (We will assume that all sets of measure zero are included in $\mathcal{F}_t$).

One often thinks of $\mathcal{F}_t$ as "the history of B_s up to time t". A function $h(\omega)$ will be $\mathcal{F}_t$-measurable if and only if h can be written as the pointwise a.e. limit of sums of functions of the form

$$g_1(B_{t_1})\, g_2(B_{t_2}) \cdots g_k(B_{t_k}),$$

where $g_1,\ldots,g_k$ are Borel functions.
Intuitively, that h is $_t$-measurable means that the value of $h(\omega)$ can be decided from the values of $B_s(\omega)$ for $s \le t$. For example,
$h_1(\omega) = B_{t/2}(\omega)$ is $\mathcal{F}_t$-measurable, while
$h_2(\omega) = B_{2t}(\omega)$ is not.

Note that $\mathcal{F}_s \subset \mathcal{F}_t$ for $s<t$ (i.e. $\{\mathcal{F}_t\}$ is non-decreasing) and that $\mathcal{F}_t \subset \mathcal{F}$ for all t.

<u>DEFINITION 3.3</u> Let $\{\mathcal{N}_t\}_{t\ge 0}$ be a non-decreasing family of σ-algebras of subsets of Ω. A process $g(t,\omega)\colon [0,\infty) \times \Omega \to \mathbb{R}^n$ is called <u>$\mathcal{N}_t$-adapted</u> if for each $t\ge 0$ the function

$$\omega \to g(t,\omega)$$

is $\mathcal{N}_t$-measurable.

Thus the process $h_1(t,\omega) = B_{t/2}(\omega)$ is $\mathcal{F}_t$-adapted, while $h_2(t,\omega) = B_{2t}(\omega)$ is not.

We now describe our class of functions for which the Ito integral will be defined:

<u>DEFINITION 3.4</u> Let $N=N(S,T)$ be the class of functions

$$f(t,\omega)\colon[0,\infty) \times \Omega \to \mathbb{R}$$

such that

(i) $(t,\omega) \to f(t,\omega)$ is $\mathcal{B}\times\mathcal{F}$-measurable, where $\mathcal{B}$ denotes the Borel σ-algebra on $[0,\infty)$.

(ii) $f(t,\omega)$ if $\mathcal{F}_t$-adapted

(iii) $E[\int_S^T f(t,\omega)^2 dt] < \infty$

<u>The Ito integral</u>

For functions $f \in N$ we will now show how to define the <u>Ito integral</u>

$$\mathcal{I}[f](\omega) = \int_S^T f(t,\omega)dB_t(\omega),$$

where B_t is 1-dimensional Brownian motion.

The idea is natural: First we define $\mathcal{I}[\phi]$ for a simple class of functions ϕ. Then we show that each $f \in N$ can be approximated (in an appropriate sense) by such ϕ's and we use this to define $\int f dB$

as the limit of $\int \phi dB$ as $\phi \to f$. We now give the details of this construction:

A function $\phi \in N$ is called <u>elementary</u> if it has the form

(3.9) $$\phi(t,\omega) = \sum_j e_j(\omega) \cdot \chi_{[t_j,t_{j+1})}(t)$$

Note that since $\phi \in N$ each function e_j must be $\mathcal{F}_{t_j}$-measurable. Thus in Example 1 above the function ϕ_1 elementary while ϕ_2 is not.

For elementary functions $\phi(t,\omega)$ we define the integral according to (3.8), i.e.

(3.10) $$\int_S^T \phi(t,\omega)dB_t(\omega) = \sum_{j\geq 0} e_j(\omega)[B_{t_{j+1}} - B_{t_j}](\omega).$$

Now we make the following important observation:

<u>LEMMA 3.5</u> (The basic isometry). If $\phi(t,\omega)$ is bounded and elementary then

(3.11) $$E[(\int_S^T \phi(t,\omega)dB_t(\omega))^2] = E[\int_S^T \phi(t,\omega)^2 dt].$$

<u>Proof of (3.11)</u>: Put $\Delta B_j = B_{t_{j+1}} - B_{t_j}$. Then

$$E[e_i e_j \Delta B_i \Delta B_j] = \begin{cases} 0 & \text{if } i \neq j \\ E[e_j^2]\cdot(t_{j+1}-t_j) & \text{if } i = j, \end{cases}$$

using that $e_i e_j \Delta B_i$ and ΔB_j are independent if $i < j$. Thus

$$E[(\int_S^T \phi dB)^2] = \sum_{i,j} E[e_i e_j \Delta B_i \Delta B_j] = \sum_j E[e_j^2]\cdot(t_{j+1}-t_j)$$

$$= E[\int_S^T \phi^2 dt].$$

The idea is now to use the isometry (3.11) to extend the definition for elementary functions to functions in N. We do this in several steps:

<u>STEP 1</u>. Let $g \in N$ be bounded and $g(\cdot,\omega)$ continuous for each ω. Then there exist elementary functions $\phi_n \in N$ such that

$$E[\int_S^T (g-\phi_n)^2 dt] \to 0 \quad \text{as } n \to \infty.$$

Proof of Step 1. Define $\phi_n(t,\omega) = \sum_j g(t_j,\omega)\cdot\chi_{[t_j,t_{j+1})}(t)$

Then ϕ_n is elementary since $g \in N$, and

$$\int_S^T (g-\phi_n)^2 dt \to 0 \quad \text{as} \quad n \to \infty, \quad \text{for each } \omega,$$

since $g(\cdot,\omega)$ is continuous for each ω.

Hence $E[\int_S^T (g-\phi_n)^2 dt] \to 0$ as $n \to \infty$, by bounded convergence.

STEP 2. Let $h \in N$ be bounded. Then there exist bounded functions $g_n \in N$ such that $g_n(\cdot,\omega)$ is continuous, for all ω and n, and

$$E[\int_S^T (h-g_n)^2 dt] \to 0.$$

Proof. Suppose $|h(t,\omega)| \leq M$ for all (t,ω). For each n let ψ_n be a non-negative, continuous function on $\mathbb{R}$ such that

(i) $\psi_n(x) = 0$ for $x \leq -\frac{1}{n}$ and $x \geq 0$

(ii) $\int_{-\infty}^{\infty} \psi_n(x)dx = 1$

Define

$$g_n(t,\omega) = \int_0^t \psi_n(s-t)h(s,\omega)ds.$$

Then $g_n(\cdot,\omega)$ is continuous for each ω and $|g_n(t,\omega)| \leq M$.
Since $h \in N$ we see that $g_n(t,\cdot)$ is $\mathcal{F}_t$-measurable for all t.
(use sums to approximate the integral defining g_n).
Moreover,

$$\int_S^T (g_n(s,\omega)-h(s,\omega))^2 ds \to 0 \quad \text{as} \quad n \to \infty, \quad \text{for each } \omega,$$

since $\{\psi_n\}_n$ constitute an approximate identity. (See Hoffman [1], p. 22).
So by bounded convergence

$$E[\int_S^T (h(t,\omega)-g_n(t,\omega))^2 dt] \to 0 \quad \text{as} \quad n \to \infty,$$

as asserted.

STEP 3. Let $f \in N$. Then there exists a sequence $\{h_n\} \subset N$ such that h_n is bounded for each n and

$$E[\int_S^T (f-h_n)^2 dt] \to 0 \quad \text{as} \quad n \to \infty.$$

Proof. Put $h_n(t,\omega) = \begin{cases} -n & \text{if } f(t,\omega) < -n \\ f(t,\omega) & \text{if } -n \le f(t,\omega) \le n \\ n & \text{if } f(t,\omega) > n \end{cases}$

Then the conclusion follows by dominated convergence.

That completes the approximation procedure.

We are now ready to complete the definition of the Ito integral

$$\int_S^T f(t,\omega)dB_t(\omega) \; ; \quad \text{for } f \in N.$$

If $f \in N$ we choose, by Steps 1-3, elementary functions $\phi_n \in N$ such that

$$E[\int_S^T |f-\phi_n|^2 dt] \to 0.$$

Then define

$$(3.12) \qquad \mathscr{I}[f](\omega) = \int_S^T f(t,\omega)dB_t(\omega) = \lim_{n\to\infty} \int_S^T \phi_n(t,\omega)dB_t(\omega).$$

The limit exists as an element of $L^2(\Omega,P)$, since $\{\int_S^T \phi_n(t,\omega)dB_t(\omega)\}$ forms a Cauchy sequence in $L^2(\Omega,P)$, by (3.11). Also by (3.11) the limit is independent of the sequence ϕ_n, and

$$(3.13) \qquad E[(\int_S^T f(t,\omega)dB_t)^2] = E[\int_S^T f^2(t,\omega)dt]. \qquad \text{(The Ito isometry)}$$

We illustrate this integral with an example:

EXAMPLE 3.6.

$$\int_0^t B_s dB_s = \frac{1}{2} B_t^2 - \frac{1}{2} t.$$

Proof. Put $\phi_n(s,\omega) = \Sigma\, B_j(\omega) \cdot \chi_{[t_j,t_{j+1})}(s)$, where $B_j = B_{t_j}$.

Then

$$E[\int_0^t (\phi_n - B_s)^2 ds] = E[\sum_j \int_{t_j}^{t_{j+1}} (B_j-B_s)^2 ds]$$

$$= \sum_j \int_{t_j}^{t_{j+1}} (s-t_j)ds = \sum_j \frac{1}{2}(t_{j+1}-t_j)^2 \to 0 \quad \text{as} \quad \Delta t_j \to 0.$$

So

$$\int_0^t B_s dB_s = \lim_{\Delta t_j \to 0} \int_0^t \phi_n dB_s .$$

Now

$$E[(\tfrac{1}{2} B_t^2 - \tfrac{1}{2} t - \Sigma B_j \Delta B_j)^2]$$

$$= E[\tfrac{1}{4} B_t^4 + \tfrac{1}{4} t^2 + (\Sigma B_j \Delta B_j)^2 - B_t^2 \cdot \Sigma B_j \Delta B_j - \tfrac{1}{2} t B_t^2 + t \cdot \Sigma B_j \Delta B_j]$$

$$= \tfrac{3}{4} t^2 + \tfrac{1}{4} t^2 + E[(\Sigma B_j \Delta B_j)^2] - E[B_t^2 \cdot \Sigma B_j \Delta B_j] - \tfrac{1}{2} t^2 + 0.$$

We have

$$E[(\Sigma B_j \Delta B_j)^2] = \sum_{i,j} E[B_i B_j \Delta B_i \Delta B_j] = \sum_i E[B_i^2 (\Delta B_i)^2] = \sum_i t_i \Delta t_i$$

and

$$E[B_t^2 \cdot \Sigma B_j \Delta B_j] = \sum_j E[B_t^2 B_j \Delta B_j] = \sum_j E[\{(B_t - B_{j+1}) + \Delta B_j + B_j\}^2 B_j \cdot \Delta B_j]$$

$$= \sum_j E[(B_t - B_{j+1})^2 B_j \Delta B_j] + \sum_j E[B_j \cdot (\Delta B_j)^3] + \sum_j E[B_j^3 \cdot \Delta B_j]$$

$$+ 2\Sigma E[(B_t - B_{j+1}) B_j \Delta B_j^2] + 2 \cdot \sum_j E[(B_t - B_{j+1}) B_j^2 \Delta B_j + 2 \sum_j E[B_j^2 \Delta B_j^2]$$

$$= 0 + 0 + 0 + 0 + 0 + 2 \sum_j t_j \cdot \Delta t_j$$

Hence $E[(\tfrac{1}{2} B_t^2 - \tfrac{1}{2} t - \Sigma B_j \Delta B_j)^2]$

$= \tfrac{1}{2} t^2 - \sum_j t_j \Delta t_j \to 0$ as $\Delta t_j \to 0$!

The extra term $-\tfrac{1}{2} t$ shows that the Ito stochastic integral does not behave like ordinary integrals. In the next chapter we will establish the Ito formula, which explains the result in this example and which makes is easy to calculate many stochastic integrals.

<u>Some properties of the Ito integral</u>

First we observe the following:

<u>THEOREM 3.7.</u> Let $f, g \in N(0,T)$ and let $0 \leq S < U < T$. Then

(i) $\int_S^T f dB_t = \int_S^U f dB_t + \int_U^T f dB_t$ for a.a.ω

(ii) $\int_S^T (cf+g) dB_t = c \cdot \int_S^T f dB_t + \int_S^T g dB_t$ (c constant) for a.a.ω

(iii) $E[\int_S^T f dB_t] = 0$

Proof. This clearly holds for all elementary functions, so by taking limits we obtain this for all $f, g \in N(0,T)$.

An important property of the Ito integral is that it is a martingale:

DEFINITION 3.8. A stochastic process $\{M_t\}_{t\geq 0}$ on $(\Omega,\mathcal{F},P)$ is called a martingale with respect to a non-decreasing sequence $\{\mathcal{M}_t\}_{t\geq 0}$ of σ-algebras if

(i) M_t is $\mathcal{M}_t$-measurable, for all t,

(ii) $E[|M_t|] < \infty$ for all t

and

(iii) $E[M_s|\mathcal{M}_t] = M_t$ for all $s \geq t$.

(See Appendix B for a survey of conditional expectation).

EXAMPLE 3.9. Brownian motion B_t is a martingale w.r.t. the σ-algebra $\mathcal{F}_t$ generated by $\{B_s\ ;\ s \leq t\}$, because

$$E[|B_t|]^2 \leq 1 \cdot E[|B_t|^2] = |B_0|^2 + t \quad \text{and}$$

$$E[B_s|\mathcal{F}_t] = E[B_s - B_t + B_t|\mathcal{F}_t]$$

$$= E[B_s - B_t|\mathcal{F}_t] + E[B_t|\mathcal{F}_t] = 0 + B_t = B_t.$$

For continuous martingales we have the following important inequality due to Doob: See e.g. Stroock and Varadhan [1], th. 1.2.3).

THEOREM 3.10. (Doob's martingale inequality) If M_t is a martingale such that $t \to M_t(\omega)$ is continuous, a.s., then for all $p\geq 1$ and all $\lambda > 0$

$$P[\sup_{0\leq t\leq T} |M_t| \geq \lambda] \leq \frac{1}{\lambda^p} \cdot E[|M_T|^p].$$

We now use this inequality to prove that the Ito integral

$$\int_0^t f(s,\omega)dB_s$$

can be chosen to depend continuously on t:

THEOREM 3.11. Let $f \in N(0,T)$. Then there exists a t-continuous version of

$$\int_0^t f(s,\omega)dB_s(\omega)\ ;\ 0 \leq t \leq T,$$

i.e. there exists a t-continuous stochastic process J_t on $(\Omega,\mathcal{F},P)$ such that

$$\text{(3.14)}\qquad P[J_t = \int_0^t f dB] = 1 \quad \text{for all } t,\ 0 \leq t \leq T.$$

Proof. Let ϕ_n be elementary functions such that

$$E[\int_0^T (f-\phi_n)^2 dt] \to 0 \quad \text{when } n \to \infty.$$

Put

$$\text{(3.15)}\qquad I_n(t,\omega) = \int_0^t \phi_n(s,\omega)dB_s(\omega) \quad \text{and}$$

$$I_t = I(t,\omega) = \int_0^t f(s,\omega)dB_s(\omega)\ ;\ 0 \leq t \leq T.$$

Then $I_n(\cdot,\omega)$ is continuous, for all n. Moreover, $I_n(t,\omega)$ is a martingale with respect to $\mathcal{F}_t$, for all n:

$$\text{(3.16)}\qquad E[I_n(t,\omega)|\mathcal{F}_s] = E[\int_0^s \phi_n dB + \int_s^t \phi_n dB|\mathcal{F}_s]$$

$$= \int_0^s \phi_n dB = I_n(s,\omega) \quad \text{when } s < t.$$

Hence $I_n - I_m$ is also an $\mathcal{F}_t$-martingale, so by the martingale inequality (Theorem 3.10) it follows that

$$P[\sup_{0\leq t\leq T} |I_n(t,\omega) - I_m(t,\omega)| > \varepsilon] \leq \frac{1}{\varepsilon^2} \cdot E[|I_n(T,\omega) - I_m(T,\omega)|^2]$$

$$= \frac{1}{\varepsilon^2} E[\int_0^T (\phi_n-\phi_m)^2 ds] \to 0 \quad \text{as } m,n \to \infty.$$

Hence we may choose a subsequence $n_k \uparrow \infty$ s.t.

$$P[\sup_{0\leq t\leq T} |I_{n_{k+1}}(t,\omega) - I_{n_k}(t,\omega)| > 2^{-k}] < 2^{-k}.$$

By the Borel-Cantelli lemma

$$P[\sup_{0\leq t\leq T} |I_{n_{k+1}}(t,\omega) - I_{n_k}(t,\omega)| > 2^{-k} \text{ for infinitely many } k] = 0.$$

So for a.a. ω there exists $k_1(\omega)$ such that

$$\sup_{0\leq k\leq T} |I_{n_{k+1}}(t,\omega) - I_{n_k}(t,\omega)| < 2^{-k} \quad \text{for } k > k_1(\omega).$$

Therefore $I_{n_k}(t,\omega)$ is uniformly convergent for $t\in[0,T]$, for a.a.ω and so the limit, denoted by $J_t(\omega)$, is t-continuous for $t \in [0,T]$, a.s. Since $I_{n_k}(t,\cdot) \to I(t,\cdot)$ in $L^2[P]$ for all t, we must have

$$I_t = J_t \quad \text{a.s., for all } t \in [0,T].$$

That completes the proof.

From now on we shall always assume that $\int_0^t f(s,\omega)dB_s(\omega)$ means a t-continuous version of the integral.

<u>COROLLARY 3.12</u>. Let $f(t,\omega) \in N(0,\infty)$.
Then

$$M_t(\omega) = \int_0^t f(s,\omega)dB_s$$

is a martingale wrt. $\mathcal{F}_t$ and

$$(3.17) \qquad P[\sup_{0\le t\le T} |M_t| \ge \lambda] \le \frac{1}{\lambda^2} \cdot E[\int_0^T f(s,\omega)^2 ds] \;;\; \lambda, T > 0.$$

<u>Proof</u>. This follows from (3.16), the a.s. t-continuity of M_t and the martingale inequality (Theorem 3.10), combined with the Ito isometry (3.13).

<u>Extensions of the Ito integral</u>

The Ito integral $\int fdB$ can be defined for a larger class of integrands f than N. First, the measurability condition (ii) of Definition 3.4 can be relaxed to the following:

(ii)'. There exists a non-decreasing family of σ-algebras $\mathcal{H}_t; t\ge 0$ such that

a) B_t is a martingale with respect to $\mathcal{H}_t$ and

b) f_t is $\mathcal{H}_t$ - adapted

Note that a) implies that $\mathcal{F}_t \subseteq \mathcal{H}_t$. The essence of this extension is that we can allow f_t to depend on more than $\mathcal{F}_t$ as long as B_t remains a martingale with respect to the "history" of $f_s; s\le t$. If (ii)' holds, then $E[B_s - B_t | \mathcal{H}_t] = 0$ for all $s > t$ and if we inspect our proofs above, we see that this is sufficient to carry out the construction of the Ito integral as before.

The most important example of a situation where (ii)' applies (and (ii) doesn't) is the following:

Suppose $B_t(\omega) = B_k(t,\omega)$ is the k'th coordinate of n-dimensional Brownian motion $(B_1,\dots,B_n)$. Let $\mathcal{F}_t^{(n)}$ be the σ-algebra generated by $B_1(s,\cdot),\dots,B_n(s,\cdot)$; $s \le t$. Then $B_k(t,\omega)$ is a martingale with respect to $\mathcal{F}_t^{(n)}$ because $B_k(s,\cdot)-B_k(t,\cdot)$ is independent of $\mathcal{F}_t^{(n)}$ when $s > t$. Thus we have now defined $\int_0^t f(s,\omega)dB_k(s,\omega)$ for $\mathcal{F}_t^{(n)}$-adapted integrands f_t. That includes integrals like

$$\int B_2 dB_1 \qquad \text{or} \qquad \int (B_1^2+B_2^2)\ dB_2$$

involving several components of n-dimensional Brownian motion. (Here we have used the notation $dB_1 = dB_1(t,\omega)$ etc.)

This allows us to define the <u>multi-dimensional Ito integral</u> as follows:

Let $B=(B_1,B_2,\dots,B_n)$ be n-dimensional Brownian motion and let $V=[V_{ij}(t,\omega)]$ be an $m\times n$ matrix where each entry $V_{ij}(t,\omega)$ satisfies (i) and (iii) of Definition 3.4 and is $\mathcal{F}_t^{(n)}$-adapted (or more generally satisfies (ii)' with respect to all B_k). Then, using matrix notation, we define

$$\int_0^t VdB = \int_0^t \begin{bmatrix} V_{11}, \cdots & V_{1n} \\ \vdots & \\ V_{m1} & \cdots\ V_{mn} \end{bmatrix} \begin{bmatrix} dB_1 \\ \vdots \\ dB_n \end{bmatrix}$$

to be the $m\times 1$ matrix (column vector) whose i'th component is the following sum of 1-dimensional Ito integrals:

$$\sum_{j=1}^{n} \int_0^t V_{ij}(s,\omega)dB_j(s,\omega)$$

The next extension of the Ito integral consists of weakening condition (iii) of Definition 3.4 to

$$\text{(iii)'} \qquad P[\int_0^t f(s,\omega)^2 ds < \infty \quad \text{for all} \quad t \ge 0] = 1$$

<u>DEFINITION 3.13</u>. M denotes the class of processes satisfying (i) of Definition 3.4 and (ii)', (iii)' above.

Let B_t denote 1-dimensional Brownian motion.
If $f \in M$ one can show that for all t there exist step functions $f_n \in N[0,t]$ such that $\int_0^t |f_n-f|^2 ds \to 0$ in probability, i.e. in measure with respect to P. For such a sequence one has that

$\int_0^t f_n(s,\omega)dB_s$ converges in probability to some random variable and the limit only depend on f, not on the sequence $\{f_n\}$. Thus we may define

$$(3.18) \qquad \int_0^t f(s,\omega)dB_s(\omega) = \text{P-}\lim_{n\to\infty} \int_0^t f_n(s,\omega)dB_s(\omega) \quad \text{for} \quad f \in M.$$

As before there exists a t-continuous version of this integral. See Friedman [1] Ch. 4 or McKean [1], Ch. 2 for details.

<u>A comparison of Ito and Stratonovich integrals</u>

Let us now return to our original question in this chapter: We have argued that the mathematical interpretation of the white noise equation

$$(3.2) \qquad \frac{dX}{dt} = b(t,X_t) + \sigma(t,X_t) \cdot W_t$$

is that X_t is a solution of the integral equation

$$(3.5) \qquad X_t = X_0 + \int_0^t b(s,X_s)ds + \int_0^t \sigma(s,X_s)dB_s,$$

for some suitable interpretation of the last integral in (3.5). However, as indicated earlier, the Ito interpretation of an integral of the form

$$(*) \qquad \int_0^t f(s,\omega)dB_s(\omega)$$

is just one of several reasonable choices. For example, the Stratonovich integral is another possibility, leading (in general) to a different result. So the question still remains: <u>Which interpretation of (*) makes (3.5) the "right" mathematical model for the equation (3.2)?</u>

Here is an argument that indicates that the <u>Stratonovich</u> interpretation in some situations may be the most appropriate: Choose t-continuously differentiable processes $B_t^{(n)}$ such that for a.a.ω

$$B^{(n)}(t,\omega) \to B(t,\omega) \quad \text{as} \quad n \to \infty$$

uniformly (in t) in bounded intervals. For each ω let $X_t^{(n)}(\omega)$ be the solution of the corresponding (deterministic) differential equation

$$(3.19) \qquad \frac{dX_t}{dt} = b(t,X_t) + \sigma(t,X_t)\frac{dB_t^{(n)}}{dt}$$

Then $X_t^{(n)}(\omega)$ converges to some function $X_t(\omega)$ in the same sense:

For a.s.ω we have that $X_t^{(n)}(\omega) \to X_t(\omega)$ as $n \to \infty$, uniformly (in t) in bounded intervals. It turns out that this solution X_t coincides with the solution of (3.5) obtained by using Stratonovich integrals. (See Wong and Zakai [1] and Sussman [1]). This implies that X_t is the solution of the following modified Ito equation:

$$(3.20) \qquad X_t = X_0 + \int_0^t b(s,X_s)ds + \int_0^t \tfrac{1}{2}\sigma'(s,X_s)\sigma(s,X_s)ds + \int_0^t \sigma(s,X_s)dB_s,$$

where σ' denotes the derivative of $\sigma(t,x)$ with respect to x. (See Stratonovich [1]).

Therefore, expressed with Ito integrals it seems reasonable from this point of view to use (3.20) (i.e. the Stratonovich interpretation) - and not (3.5) (the Ito interpretation) - as the model for the original white noise equation (3.2).

On the other hand, the specific feature of the Ito model of "not looking into the future" (as explained after Example 3.1) seems to be a reason for choosing the Ito interpretation in many cases, for example in biology (see the discussion in Turelli [1]). The difference between the two interpretations is illustrated in Example 5.1. Note that (3.20) and (3.5) coincide if $\sigma(t,x)$ does not depend on x. For example, this is the situation in the linear case handled in the filtering problem in Chapter VI.

In any case, because of the explicit connection (3.20) between the two models (and a similar connection in higher dimensions - see (6.2)), it will for many purposes suffice to do the general mathematical treatment for one of the two types of integrals. In general one can say that the Stratonovich integral has the advantage of leading to ordinary chain rule formulas under a transformation (change of variable), i.e. there are no 2nd order terms in the Stratonovich analogue of the Ito transformation formula (see Theorems 4.2 and 4.6). This property makes the Stratonovich integral natural to use for example in connection with stochastic differential equations on manifolds (see Elworthy [1] or Ikeda and Watanabe [1]).
However, Stratonovich integrals are not martingales, as we have seen that Ito integrals are. This gives the Ito integral an important computational advantage, even though it does not behave so nicely under transformations (as the preceding Example 3.6 shows). For our purposes the Ito integral will be most convenient, so we will base our discussion on that from now on.

IV. Stochastic Integrals and the Ito Formula

The 1-dimensional Ito formula

Example 3.6 illustrates that the basic definition of Ito integrals is not very useful when we try to evaluate a given integral. This is similar to the situation for ordinary Riemann integrals, where we do not use the basic definition but rather the fundamental theorem of calculus plus the chain rule in the explicit calculations.

In this context, however, we have no differentiation theory, only integration theory. Nevertheless it turns out that it is possible to establish an Ito integral version of the chain rule, called the Ito formula. The Ito formula is, as we will show by examples, very useful for evaluating Ito integrals.

From the example

$$\int_0^t B_s dB_s = \frac{1}{2} B_t^2 - \frac{1}{2} t$$

or

$$\frac{1}{2} B_t^2 = \frac{1}{2} t + \int_0^t B_s dB_s, \tag{4.1}$$

we see that the image of the Ito integral $B_t = \int_0^t dB_s$ by the map $g(x) = \frac{1}{2} x^2$ is not again an Ito integral of the form

$$\int_0^t f(s,\omega) dB_s(\omega)$$

but a combination of a dB_s- and a ds-integral:

$$\frac{1}{2} B_t^2 = \int_0^t \frac{1}{2} ds + \int_0^t B_s dB_s. \tag{4.2}$$

It turns out that if we introduce stochastic integrals as a sum of a dB_s- and a ds-integral then this family of integrals is stable under smooth maps. Thus we define

DEFINITION 4.1. Let B_t be 1-dimensional Brownian motion on $(\Omega, \mathcal{F}, P)$.

A (1-dimensional) stochastic integral is a stochastic process X_t on $(\Omega, \mathcal{F}, P)$ of the form

$$X_t = X_0 + \int_0^t u(s,\omega) ds + \int_0^t v(s,\omega) dB_s, \tag{4.3}$$

where $v \in M$, so that

(4.4) $$P[\int_0^t v(s,\omega)^2 ds < \infty \quad \text{for all } t \geq 0] = 1$$

(see Definition 3.13). We also assume that u is $\mathcal{H}_t$-adapted (where $\mathcal{H}_t$ is as in (ii)', Definition 3.13) and

(4.5) $$P[\int_0^t |u(s,\omega)| ds < \infty \quad \text{for all } t \geq 0] = 1.$$

If X_t is a stochastic integral of the form (4.3) the equation (4.3) is sometimes written in the shorter differential form

(4.6) $$dX_t = u\,dt + v\,dB_t.$$

For example, (4.1) (or (4.2)) may be represented by

$$d(\tfrac{1}{2} B_t^2) = \tfrac{1}{2} dt + B_t dB_t.$$

We are now ready to state the first main result in this chapter:

THEOREM 4.2. (THE 1-DIMENSIONAL ITO FORMULA). Let X_t be a stochastic integral:

$$dX_t = u\,dt + v\,dB_t.$$

Let $(t,x) \to g(t,x) \in \mathbb{R}$; $t \in [0,\infty)$, $x \in \mathbb{R}$ be C^2 (i.e. twice continuously differentiable) on $[0,\infty) \times \mathbb{R}$. Then

$$Y_t = g(t,X_t)$$

is again a stochastic integral, and

(4.7) $$dY_t = \frac{\partial g}{\partial t}(t,X_t)dt + \frac{\partial g}{\partial x}(t,X_t)dX_t + \frac{1}{2}\frac{\partial^2 g}{\partial x^2}(t,X_t) \cdot (dX_t)^2,$$

where

(4.8) $$dt \cdot dt = dt \cdot dB_t = dB_t \cdot dt = 0, \quad dB_t \cdot dB_t = dt.$$

Before we prove Ito's formula let us look at some examples.

EXAMPLE 4.3. Let us return to the integral

$$I = \int_0^t B_s dB_s \quad \text{from Chapter III.}$$

Choose $X_t = B_t$ and $g(t,x) = \frac{1}{2} x^2$.

Then

$$Y_t = g(t,B_t) = \tfrac{1}{2} B_t^2.$$

Then, by Ito's formula,

$$dY_t = \frac{\partial g}{\partial t} \cdot dt + \frac{\partial g}{\partial x} \cdot dB_t + \frac{1}{2} \cdot \frac{\partial^2 g}{\partial x^2} \cdot (dB_t)^2$$

$$= 0 + B_t \cdot dB_t + \frac{1}{2} \cdot (dB_t)^2$$

$$= B_t \cdot dB_t + \frac{1}{2} \cdot dt.$$

Hence

$$d(\frac{1}{2} B_t^2) = B_t \cdot dB_t + \frac{1}{2} dt.$$

In other words,

$$\frac{1}{2} B_t^2 = \int_0^t B_s dB_s + \frac{1}{2} t, \quad \text{as in Chapter III.}$$

EXAMPLE 4.4. What is

$$\int_0^t s \, dB_s ?$$

From classical calculus it seems reasonable that a term of the form tB_t should appear, so we put

$$g(t,x) = t \cdot x$$

and

$$Y_t = g(t,B_t) = t \cdot B_t.$$

Then by Ito's formula,

$$dY_t = B_t dt + t \, dB_t + 0$$

i.e.

$$d(tB_t) = B_t \cdot dt + t \, dB_t$$

or

$$tB_t = \int_0^t B_s ds + \int_0^t s \, dB_s$$

or

$$\int_0^t s \, dB_s = tB_t - \int_0^t B_s ds,$$

which is reasonable from an integration-by-parts point of view.

More generally, the same method gives

THEOREM 4.5. (INTEGRATION BY PARTS). Suppose $f(s,\omega) = f(s)$ only depends on s and that f is of bounded variation in $[0,t]$. Then

$$\int_0^t f(s)dB_s = f(t)B_t - \int_0^t B_s df_s$$

Note that it is crucial for the result to hold that f does not depend on ω.

Proof of the Ito formula. First observe that if we substitute

$$dX_t = u\, dt + v\, dB_t$$

and use (4.8) we get the equivalent expression

$$g(t,X_t) = g(0,X_0) + \int_0^t \left(\frac{\partial g}{\partial s} + u \cdot \frac{\partial g}{\partial x} + \frac{1}{2} v^2 \cdot \frac{\partial^2 g}{\partial x^2} \right) ds \tag{4.9}$$

$$+ \int_0^t v \cdot \frac{\partial g}{\partial x} \cdot dB_s .$$

Note that (4.9) is a stochastic integral in the sense of Definition 4.1.

We may assume that g, $\frac{\partial g}{\partial t}$, $\frac{\partial g}{\partial x}$ and $\frac{\partial^2 g}{\partial x^2}$ are bounded, for if (4.9) is proved in this case we obtain the general case by approximating by C^2 functions g_n such that g_n, $\frac{\partial g_n}{\partial t}$, $\frac{\partial g_n}{\partial x}$ and $\frac{\partial^2 g_n}{\partial x^2}$ are bounded for each n and converge uniformly on compact subsets of $[o,\infty) \times \mathbb{R}$ to g, $\frac{\partial g}{\partial t}$, $\frac{\partial g}{\partial x}$, $\frac{\partial^2 g}{\partial x^2}$, respectively. Using Taylor's theorem we get

$$g(t,X_t) = g(0,X_0) + \sum_j \Delta g(t_j,X_j) = g(0,X_0) + \sum_j \frac{\partial g}{\partial t} \Delta t_j + \sum_j \frac{\partial g}{\partial x} \cdot \Delta X_j$$

$$+ \frac{1}{2} \sum_j \frac{\partial^2 g}{\partial t^2} \cdot (\Delta t_j)^2 + \sum_j \frac{\partial^2 g}{\partial t \partial x} \cdot (\Delta t_j)(\Delta X_j) + \frac{1}{2} \cdot \sum_j \frac{\partial^2 g}{\partial x^2} \cdot (\Delta X_j)^2 + \sum R_j ,$$

where $\frac{\partial g}{\partial t}$, $\frac{\partial g}{\partial x}$ etc. are evaluated at the points (t_j, X_{t_j}),

$\Delta t_j = t_{j+1} - t_j$, $\Delta X_j = X_{t_{j+1}} - X_{t_j}$, $\Delta g(t_j,X_j) = g(t_{j+1},X_{t_{j+1}}) - g(t_j,X_j)$

and $R_j = o(|\Delta t_j|^2 + |\Delta X_j|^2)$ for all j.

If $\Delta t_j \to 0$ then

$$(4.10)\qquad \sum_j \frac{\partial g}{\partial t}\cdot \Delta t_j = \sum_j \frac{\partial g}{\partial t}(t_j,X_j)\Delta t_j \to \int_0^t \frac{\partial g}{\partial t}(s,X_s)ds$$

$$(4.11)\qquad \sum_j \frac{\partial g}{\partial x}\Delta X_j = \sum_j \frac{\partial g}{\partial x}(t_j,X_j)\Delta X_j \to \int_0^t \frac{\partial g}{\partial x}(s,X_s)dX_s.$$

Moreover,

$$(4.12)\qquad \sum_j \frac{\partial^2 g}{\partial x^2}(\Delta X_j)^2 = \sum_j \frac{\partial^2 g}{\partial x^2} u_j^2(\Delta t_j)^2 + 2\sum_j \frac{\partial^2 g}{\partial x^2} u_j v_j(\Delta t_j)(\Delta B_j)$$

$$+ \sum_j \frac{\partial^2 g}{\partial x^2} v_j^2 \cdot (\Delta B_j)^2.$$

The first two terms here tend to 0 as $\Delta t_j \to 0$.

(For example, $E[(\sum_j \frac{\partial^2 g}{\partial x^2} u_j v_j(\Delta t_j)(\Delta B_j))^2]$

$$= \sum E[(\frac{\partial^2 g}{\partial x^2} u_j v_j)^2](\Delta t_j)^3 \to 0 \quad \text{as} \quad \Delta t_j \to 0).$$

We claim that the last term tends to

$$\int_0^t \frac{\partial^2 g}{\partial x^2} v^2 ds \quad \text{in} \quad L^2(\Omega), \quad \text{as} \quad \Delta t_j \to 0$$

To prove this, put $a(t) = \frac{\partial^2 g}{\partial x^2}(t,X_t)v^2(t,\omega)$, $a_j = a(t_j)$ and consider

$$E[(\sum_j a_j(\Delta B_j)^2 - \sum_j a_j \Delta t_j)^2]$$

$$= \sum_{i,j} E[a_i a_j((\Delta B_i)^2 - \Delta t_i)((\Delta B_j)^2 - \Delta t_j)]$$

If $i < j$ then $a_i a_j((\Delta B_i)^2 - \Delta t_i)$ and $(\Delta B_j)^2 - \Delta t_j$ are independent so the terms vanish in this case, and similarly if $i > j$. So we are left with

$$\sum_j E[a_j^2((\Delta B_j)^2 - \Delta t_j)^2] = \sum_j E[a_j^2]\cdot E[(\Delta B_j)^4 - 2(\Delta B_j)^2\Delta t_j + (\Delta t_j)^2]$$

$$= \sum_j E[a_j^2]\cdot(3\cdot(\Delta t_j)^2 - 2(\Delta t_j)^2 + (\Delta t_j)^2) = 2\sum_j E[a_j^2]\cdot(\Delta t_j)^2$$

$$\to 0 \quad \text{as} \quad \Delta t_j \to 0.$$

In other words, we have established that

$$\sum_j a_j(\Delta B_j)^2 \to \int_0^t a(s)ds \quad \text{in} \quad L^2(\Omega) \quad \text{as} \quad \Delta t_j \to 0$$

and this is often expressed shortly by the striking formula

(4.13) $(dB_t)^2 = dt.$

The argument above also proves that $\Sigma R_j \to 0$ as $\Delta t_j \to 0$. That completes the proof of the Ito formula.

REMARK. Note that it is enough that $g(t,x)$ is C^2 on $[0,\infty) \times U$, if $U \subset \mathbb{R}$ is an open set such that $X_t(\omega) \in U$ for all $t \geq 0$, $\omega \in \Omega$.

The multi-dimensional Ito formula

We now turn to the situation in higher dimensions: Let $B(t,\omega)=(B_1(t,\omega),\ldots,B_m(t,\omega))$ denote m-dimensional Brownian motion. If each of the processes $u_i(t,\omega)$ and $v_{ij}(t,\omega)$ satisfies the conditions given in (4.3) - (4.5) above ($1 \leq i \leq n$, $1 \leq j \leq m$) then we can form the following m stochastic integrals

$$(4.14)\qquad \begin{cases} dX_1 = u_1\,dt + v_{11}\,dB_1 + \ldots + v_{1m}\,dB_m \\ \vdots \qquad \vdots \qquad\qquad\qquad\qquad \vdots \\ dX_n = u_n\,dt + v_{n1}\,dB_1 + \ldots + v_{nm}\,dB_m \end{cases}$$

Or, in matrix notation simply

$$(4.15)\qquad dX = u\,dt + v\,dB,$$

where

$$(4.16)\qquad X = \begin{bmatrix} X_1 \\ \vdots \\ X_n \end{bmatrix},\quad u = \begin{bmatrix} u_1 \\ \vdots \\ u_n \end{bmatrix},\quad v = \begin{bmatrix} v_{11} & \cdots & v_{1m} \\ \vdots & & \vdots \\ v_{n1} & \cdots & v_{nm} \end{bmatrix},\quad dB = \begin{bmatrix} dB_1 \\ \vdots \\ dB_m \end{bmatrix}$$

We now ask: What is the result of applying a smooth function to X? The answer is given by

THEOREM 4.6. (THE GENERAL ITO FORMULA)

Let

$$dX = u\,dt + v\,dB$$

be an n-dimensional stochastic integral as above. Let $g(t,x) = (g_1(t,x),\ldots,g_p(t,x))$ be a C^2 map from $[0,\infty) \times \mathbb{R}^n$ into $\mathbb{R}^p$. Then the process

$$Y(t,\omega) = g(t,X_t)$$

is again a stochastic integral, whose component nr. k, Y_k, is given by

$$dY_k = \frac{\partial g_k}{\partial t}(t,X)dt + \sum_i \frac{\partial g_k}{\partial x_i}(t,X)dX_i + \frac{1}{2}\sum_{i,j} \frac{\partial^2 g_k}{\partial x_i \partial x_j}(t,X)dX_i dX_j$$

where $dB_i dB_j = \delta_{ij}dt$, $dB_i dt = dt dB_i = 0$.

The proof is similar to the 1-dimensional version (Theorem 4.2) and is omitted.

EXAMPLE 4.7. Let $B = (B_1,\dots,B_n)$ be Brownian motion in $\mathbb{R}^n$, $n \geq 2$, and consider

$$R(t,\omega) = |B(t,\omega)| = B_1(t,\omega)^2 + \dots + B_n^2(t,\omega),$$

i.e. the distance to the origin of $B(t,\omega)$. The function $g(t,x)=|x|$ is not C^2 at the origin, but since B_t never hits the origin, a.s. when $n \geq 2$ (see e.g. Example 7.12) Ito's formula still works and we get

$$dR = \sum_{i=1}^{n} \frac{B_i dB_i}{R} + \frac{n-1}{2R} dt$$

The process R is called the <u>n-dimensional Bessel process</u> because its generator (Chapter VII) is the Bessel differential operator $Af(x) = \frac{1}{2} f''(x) + \frac{n-1}{2x} f'(x)$. See Example 8.9.

V. Stochastic Differential Equations

We now return to the possible solutions $X_t(\omega)$ of the stochastic differential equation

(5.1) $$\frac{dX_t}{dt} = b(t,X_t) + \sigma(t,X_t)W_t, \quad b \in \mathbb{R}, \quad \sigma \in \mathbb{R}$$

where W_t is 1-dimensional "white noise". As discussed in Chapter III the Ito interpretation of (5.1) is that X_t satisfies the stochastic integral equation

(5.2) $$X_t = X_0 + \int_0^t b(s,X_s)ds + \int_0^t \sigma(s,X_s)dB_s$$

or in differential form

(5.3) $$dX_t = b(t,X_t)dt + \sigma(t,X_t)dB_t.$$

Therefore, to get from (5.1) to (5.3) we formally just replace the white noise W_t by $\frac{dB_t}{dt}$ in (5.1) and multiply by dt. It is natural to ask:

(A) Can one obtain existence and uniqueness theorems for such equations? What are the properties of the solutions?

(B) How can one solve a given such equation?

We will first consider question (B) by looking at some simple examples, and then discuss (A).

It is the Ito formula that is the key to the solution of many stochastic differential equations. The method is illustrated in the following examples.

<u>EXAMPLE 5.1</u>. Let us return to the population growth model in Chapter I:

$$\frac{dN_t}{dt} = a_t \cdot N_t, \quad N_0 \text{ given}$$

where $a_t = r_t + \alpha \cdot W_t$, W_t = white noise, α = constant.

Let us assume that $r_t = r$ = constant. By the Ito interpretation (5.3) this equation is equivalent to (here $\sigma(t,x) = \alpha x$)

$$dN_t = rN_t\, dt + \alpha N_t dB_t$$

or

$$\frac{dN_t}{N_t} = rdt + \alpha dB_t.$$

Hence

$$(5.4)\qquad \int_0^t \frac{dN_s}{N_s} = rt + \alpha B_t \qquad (B_0 = 0).$$

To evaluate the integral on the left hand side we use the Ito formula for the function

$$g(t,x) = \ln x \; ; \; x > 0$$

and obtain

$$d(\ln N_t) = \frac{1}{N_t} \cdot dN_t + \frac{1}{2}\left(-\frac{1}{N_t^2}\right)(dN_t)^2$$

$$= \frac{dN_t}{N_t} - \frac{1}{2N_t^2} \cdot \alpha^2 N_t^2 dt = \frac{dN_t}{N_t} - \frac{1}{2}\alpha^2 dt.$$

Hence

$$\frac{dN_t}{N_t} = d(\ln N_t) + \frac{1}{2}\alpha^2 dt$$

so from (5.4) we conclude

$$\ln \frac{N_t}{N_0} = (r - \frac{1}{2}\alpha^2)t + \alpha B_t$$

or

$$(5.5)\qquad \underline{N_t = N_0 \cdot \exp((r - \tfrac{1}{2}\alpha^2)t + \alpha B_t)}$$

For comparison, referring to the discussion at the end of Chapter III, the Stratonovich interpretation (3.20) would have given

$$(5.6)\qquad \bar{N}_t = N_0 \cdot \exp(rt + \alpha B_t).$$

REMARK. It seems reasonable that if B_t is independent of N_0 we should have

$$(*)\qquad E[N_t] = E[N_0]e^{rt},$$

i.e. the same as when there is no noise in a_t. To see if this is indeed the case, we let

$$Y_t = e^{\alpha B_t}$$

and apply Ito's formula:

$$dY_t = \alpha e^{\alpha B_t} dB_t + \frac{1}{2}\alpha^2 e^{\alpha B_t} \cdot dt$$

or

$$Y_t = Y_0 + \alpha \int_0^t e^{\alpha B_s} dB_s + \frac{1}{2}\alpha^2 \int_0^t e^{\alpha B_s} ds.$$

Since $E[\int_0^t e^{\alpha B_s} dB_s] = 0$ (Theorem 3.7 (iii)), we get

$$E[Y_t] = E[Y_0] + \frac{1}{2}\alpha^2 \int_0^t E[Y_s]ds$$

i.e.

$$\frac{d}{dt} E[Y_t] = \frac{1}{2}\alpha^2 \cdot E[Y_t], \quad E[Y_0] = 1.$$

So

$$E[Y_t] = e^{\frac{1}{2}\alpha^2 t},$$

and therefore - as anticipated - we obtain

$$E[N_t] = E[N_0] \cdot e^{rt}.$$

For the <u>Stratonovich</u> solution, however, the same calculation gives

$$E[\bar{N}_t] = E[N_0]\, e^{(r + \frac{1}{2}\alpha^2)t}.$$

Now that we have found the explicit solutions N_t and $\bar{N}_t$ in (5.5), (5.6) we can use our knowledge about the behaviour of B_t to gain information on these solutions. For example, for the Ito solution N_t we get the following:

(i) If $r > \frac{1}{2}\alpha^2$ then $N_t \to \infty$ as $t \to \infty$, a.s.

(ii) If $r < \frac{1}{2}\alpha^2$ then $N_t \to 0$ as $t \to \infty$, a.s.

(iii) If $r = \frac{1}{2}\alpha^2$ then N_t will fluctuate between arbitrary large and arbitrary small values as $t \to \infty$, a.s.

These conclusions are direct consequences of the formula (5.5) for N_t together with the following basic result about 1-dimensional Brownian motion B_t:

<u>THEOREM 5.2 (THE LAW OF ITERATED LOGARITM)</u>

$$\limsup_{t \to \infty} \frac{B_t}{\sqrt{2t \log \log t}} = 1 \quad \text{a.s.} \tag{5.7}$$

For a proof we refer to Lamperti [2, §22].

For the Stratonovich solution $\bar{N}_t$ we get by the same argument that $\bar{N}_t \to 0$ a.s. if $r < 0$ and $\bar{N}_t \to \infty$ a.s. if $r > 0$.
Thus the two solutions have fundamentally different properties and it is an interesting question what solution gives the best description of the situation.

EXAMPLE 5.3. Let us return to the equation in Example 2 of Chapter I:

(5.7) $$L \cdot Q_t'' + R \cdot Q_t' + \frac{1}{C} \cdot Q_t = F_t = G_t + \alpha W_t.$$

We introduce the vector

$$X = X(t,\omega) = \begin{bmatrix} X_1 \\ X_2 \end{bmatrix} = \begin{bmatrix} Q_t \\ Q_t' \end{bmatrix}$$ and obtain

(5.8) $$\begin{cases} X_1' = X_2 \\ LX_2' = -RX_2 - \frac{1}{C} \cdot X_1 + G_t + \alpha W_t \end{cases}$$

or, in matrix notation

(5.9) $$dX = AXdt + Hdt + KdB_t$$

where

(5.10) $$dX = \begin{bmatrix} dX_1 \\ dX_2 \end{bmatrix}, \quad A = \begin{bmatrix} 0 & 1 \\ -\frac{1}{CL} & -\frac{R}{L} \end{bmatrix}, \quad H_t = \begin{bmatrix} 0 \\ \frac{1}{L} G_t \end{bmatrix}, \quad K = \begin{bmatrix} 0 \\ \frac{\alpha}{L} \end{bmatrix},$$

with B_t 1-dimensional Brownian motion.

Thus we are led to a 2-dimensional stochastic differential equation. We rewrite (5.9) as

(5.11) $$\exp(-At)dX - \exp(-At)AXdt = \exp(-At)[H_t dt + KdB_t].$$

Here it is tempting to relate the left hand side to

$$d(\exp(-At)X).$$

To do this we use a 2-dimensional version of the Ito formula (Theorem 4.6).

Applying this result to the two coordinate functions g_1, g_2 of

$$g:[0,\infty) \times \mathbb{R}^2 \to \mathbb{R}^2 \text{ given by } g(t,x_1,x_2) = \exp(-At) \cdot \begin{bmatrix} x_1 \\ x_2 \end{bmatrix},$$

we obtain that

$$d(\exp(-At)X) = (-A)\exp(-At)Xdt + \exp(-At)dX.$$

Substituted in (5.11) this gives

$$\exp(-At)X - X_0 = \int_0^t \exp(-As)H_s ds + \int_0^t \exp(-As)K \cdot dB_s$$

or

$$X = \exp(At)[X_0 + \exp(-At)K \cdot B_t + \int_0^t \exp(-As)[H_s - AKB_s]ds],$$

by integration by parts (Theorem 4.5).

EXAMPLE 5.4. Choose $X = B$, 1-dimensional Brownian motion and

$$g(t,x) = e^{ix} = (\cos x, \sin x) \in \mathbb{R}^2 \quad \text{for} \quad x \in \mathbb{R}.$$

Then

$$Y = g(t,X) = e^{iB} = (\cos B, \sin B)$$

is by Ito's formula again a stochastic integral.

Its coordinates Y_1, Y_2 satisfy

$$\begin{cases} dY_1 = -\sin B \cdot dB - \frac{1}{2}\cos B \cdot dt \\ dY_2 = \cos B \cdot dB - \frac{1}{2}\sin B \cdot dt \end{cases}$$

Thus the process $Y = (Y_1, Y_2)$, which we could call Brownian motion on the unit circle, is the solution of the stochastic differential equations

$$(5.13) \qquad \begin{cases} dY_1 = -\frac{1}{2} Y_1 dt - Y_2 dB \\ dY_2 = -\frac{1}{2} Y_2 dt + Y_1 dB \end{cases}$$

Or, in matrix notation,

$$dY = -\frac{1}{2} Y dt + K \cdot Y dB, \quad \text{where} \quad K = \begin{bmatrix} 0 & -1 \\ 1 & 0 \end{bmatrix}.$$

We now turn to the existence and uniqueness question (A) above.

THEOREM 5.5. Let $T>0$ and $b(\cdot,\cdot):[0,T]\times \mathbb{R}^n \to \mathbb{R}^n, \sigma(\cdot,\cdot):[0,T]\times \mathbb{R}^n \to \mathbb{R}^{n\times m}$ be measurable functions satisfying

$$(5.14) \qquad |b(t,x)| + |\sigma(t,x)| \leq C(1+|x|) \ ; \ x \in \mathbb{R}^n, \ t \in [0,T]$$

for some constant C, (where $|\sigma|^2 = \Sigma|\sigma_{ij}|^2$) and such that

$$(5.15) \qquad |b(t,x) - b(t,y)| + |\sigma(t,x) - \sigma(t,y)| \leq D \cdot |x-y|; \ x, y \in \mathbb{R}^n, \ t \in [0,T]$$

for some constant D. Let Z be a random variable which is independent of the σ-algebra generated by $B_s(\cdot)$, $s \geq 0$ and such that

$$(5.16) \qquad E[|Z|^2] < \infty.$$

Then the stochastic differential equation

$$(5.17) \qquad dX_t = b(t,X_t)dt + \sigma(t,X_t)dB_t, \quad 0 \leq t \leq T, \quad X_0 = Z$$

has a unique t-continuous solution $X_t(\omega)$, each component of which belongs to $N[0,T]$.

<u>REMARKS</u>. Conditions (5.14) and (5.15) are natural in view of the following 2 simple examples from deterministic differential equations (i.e. $\sigma = 0$):

a) The equation

$$(5.18) \qquad \frac{dX_t}{dt} = X_t^2, \quad x(0) = 1$$

corresponding to $b(x) = x^2$ (which does not satisfy (5.14)) has the (unique) solution

$$X_t = \frac{1}{1-t}\ ; \quad 0 \leq t < 1$$

Thus it is impossible to find a global solution (defined for all t) in this case.

More generally, condition (5.14) ensures that the solution $X_t(\omega)$ of (5.17) does not <u>explode</u>, i.e. that $|X_t(\omega)|$ does not tend to ∞ in a finite time.

b) The equation

$$(5.19) \qquad \frac{dX_t}{dt} = 3\, X_t^{2/3}\ ; \quad X(0) = 0$$

has more than one solution. In fact, for any $a > 0$ the function

$$X_t = \begin{cases} 0 & \text{for } t \leq a \\ (t-a)^3 & \text{for } t > a \end{cases}$$

solves (5.19). In this case $b(x) = 3x^{2/3}$ does not satisfy the Lipschitz condition (5.15) at $x = 0$.

Thus condition (5.15) guarantees that equation (5.17) has a <u>unique</u> solution. Here uniqueness means that if $X_1(t,\omega)$ and $X_2(t,\omega)$ are two t-continuous processes in $N[0,T]$ satisfying (5.17) then

$$(5.20) \qquad X_1(t,\omega) = X_2(t,\omega) \quad \text{for all } t \leq T, \quad \text{a.s.}$$

Proof of Theorem 5.5. For simplicity we consider the case $n=m=1$. The uniqueness follows from the Ito isometry (3.13) and the Lipschitz property (5.15): Let $X_1(t,\omega) = X_t(\omega)$ and $X_2(t,\omega) = \hat{X}_t(\omega)$ be solutions with initial values Z, $\hat{Z}$ respectively, i.e. $X_1(0,\omega) = Z(\omega)$, $X_2(0,\omega) = \hat{Z}(\omega)$, $\omega \in \Omega$.

Put $a(s,\omega) = b(s,X_s) - b(s,\hat{X}_s)$ and $\gamma(s,\omega) = \sigma(s,X_s) - \sigma(s,\hat{X}_s)$.

Then
$$E[|X_t - \hat{X}_t|^2] = E[(Z-\hat{Z} + \int_0^t a\,ds + \int_0^t \gamma\,dB_s)^2]$$
$$\leq 3 \cdot E[|Z-\hat{Z}|^2] + 3 \cdot E[(\int_0^t a\,ds)^2] + 3E[(\int_0^t \gamma\,dB_s)^2]$$
$$\leq 3 \cdot E[|Z-\hat{Z}|^2] + 3t \cdot E[\int_0^t a^2 ds] + 3E[\int_0^t \gamma^2 ds]$$
$$\leq 3 \cdot E[|Z-\hat{Z}|^2] + 3\,(1+t)D^2 \cdot \int_0^t E[|X_s-\hat{X}_s|^2]ds.$$

So the function
$$v(t) = E[|X_t-\hat{X}_t|^2] \;;\; 0 \leq t \leq T$$
satisfies

$$v(t) \leq F + A \cdot \int_0^t v(s)ds, \quad \text{where} \quad F=3E[|Z-\hat{Z}|^2] \text{ and} \tag{5.21}$$

$A = 3(1+T)D^2$.

Let $w(t) = \int_0^t v(s)ds$. Then $w'(t) \leq F + A \cdot w(t)$, so since $w(0) = 0$,

$w(t) \leq \frac{F}{A}(\exp(At)-1)$. (Consider $f(t)=w(t)\exp(-At)$.)

Therefore

$$v(t) \leq F \cdot \exp(At). \tag{5.22}$$

Now assume that $Z=\hat{Z}$. Then $F=0$ and so $v(t) = 0$ for all $t \geq 0$. Hence

$$P[|X_t-\hat{X}_t| = 0 \quad \text{for all} \quad t \in \mathbb{Q} \cap [0,T]] = 1,$$

where $\mathbb{Q}$ denotes the rational numbers.

By continuity of $t \to |X_t-\hat{X}_t|$ it follows that

$$P[|X_1(t,\omega) - X_2(t,\omega)| = 0 \quad \text{for all} \quad t \in [0,T]] = 1, \tag{5.23}$$

and the uniqueness is proved.

The proof of the existence is similar to the familiar existence proof for ordinary differential equations:

Define $Y^{(0)} = X_0$ and $Y^{(k)} = Y_t^{(k)}(\omega)$ inductively as follows

$$(5.24)\qquad Y^{(k+1)} = X_0 + \int_0^t b(s,Y^{(k)})ds + \int_0^t \sigma(s,Y^{(k)})dB_s .$$

Then, similar computation as for the uniqueness above gives

$$E[|Y^{(k+1)}-Y^{(k)}|^2] \leq (1+T)3D^2 \cdot \int_0^t E[|Y^{(k)}-Y^{(k-1)}|^2]ds,$$

for $k \geq 1$

and

$$E[|Y_t^{(1)}-Y_t^{(0)}|^2] \leq 2C^2t^2(1+E[|X_0|^2])$$
$$+ 2C^2t(1+E[|X_0|^2]) \leq A_1 \cdot t$$

where the constant A_1 only depends on C, T and $E[|X_0|^2]$. So by induction on k we obtain

$$(5.25)\qquad E[|Y_t^{(k+1)}-Y_t^{(k)}|^2] \leq \frac{A_2^{k+1}t^{k+1}}{(k+1)!}\ ;\ k \geq 0,\ t \in [0,T]$$

for some suitable constant A_2 depending on C, D, T and $E[|X_0|^2]$. Now

$$\sup_{0\leq t\leq T} |Y_t^{(k+1)}-Y_t^{(k)}| \leq \int_0^T |b(s,Y_s^{(k)})-b(s,Y_s^{(k-1)})|ds$$
$$+ \sup_{0\leq t\leq T} \left|\int_0^t (\sigma(s,Y_s^{(k)}) - \sigma(s,Y_s^{(k-1)}))dB_s\right|.$$

By the martingale inequality (Theorem 3.10) we obtain

$$P[\sup_{0\leq t\leq T} |Y_t^{(k+1)}-Y_t^{(k)}| > 2^{-k}] \leq P[(\int_0^T |b(s,Y_s^{(k)})-b(s,Y_s^{(k-1)})|ds)^2 > 2^{-2k-2}]$$
$$+ P[\sup_{0\leq t\leq T} |\int_0^t (\sigma(s,Y_s^{(k)})-\sigma(s,Y_s^{(k-1)}))dB_s| > 2^{-k-1}]$$
$$\leq 2^{2k+2} \cdot T \cdot \int_0^T E(|b(s,Y_s^{(k)})-b(s,Y_s^{(k-1)})|^2)dt$$
$$+ 2^{2k+2} \cdot \int_0^T E(|\sigma(s,Y_s^{(k)})-\sigma(s,Y_s^{(k-1)})|^2)ds$$
$$\leq 2^{2k+2} \cdot D^2(T+1) \cdot \int_0^T \frac{A_2^k t^k}{k!} dt \leq \frac{(4A_2T)^{k+1}}{(k+1)!},\quad \text{if } A_2 > 4D^2(T+1).$$

Therefore, by the Borel-Cantelli lemma,

$$P[\sup_{0\le t\le T} |Y_t^{(k+1)}-Y_t^{(k)}| > 2^{-k} \quad \text{for infinitely many } k] = 0.$$

Thus, for a.a. ω there exists $k_0 = k_0(\omega)$ such that

$$\sup_{0\le t\le T} |Y_t^{(k+1)}-Y_t^{(k)}| < 2^{-k} \quad \text{for } k \ge k_0.$$

Therefore the sequence

$$Y_t^{(n)}(\omega) = Y_t^0(\omega) + \sum_{k=0}^{n-1} (Y_t^{(k+1)}(\omega)-Y_t^{(k)}(\omega))$$

is uniformly convergent in $[0,T]$, for a.a. ω.

Denote the limit by $X = X_t(\omega)$. Then X_t is t-continuous since $Y_t^{(n)}$ is t-continuous for all n. Moreover, $X_t(\cdot)$ is $\mathcal{F}_t$-measurable for all t, since $Y_t^{(n)}(\cdot)$ has this property for all n.
Next, note that for $n\ge 0$ we have

$$E[|Y_t^{(n+1)}|^2] \le 3E[|X_0|^2]+3E[(\int_0^t b(s,Y_s^{(n)})ds)^2]+3E[(\int_0^t \sigma(s,Y_s^{(n)})dB_s)^2]$$

$$\le 3E[|X_0|^2]+3C\cdot(T+1)\cdot\int_0^t E(1+|Y_s^{(n)}|^2)ds \le A_3+A_4\int_0^t E[|Y_s^{(n)}|^2]ds$$

where A_3 is a constant depending only on C, T and $E[|X_0|^2]$.

Hence by induction

$$E[|Y_t^{(n+1)}|^2] \le (1+E[|X_0|^2])\cdot\sum_{k=0}^{n+1}\frac{A_3^{k+1}t^k}{k!}.$$

So

$$(5.26) \qquad E[|X_t|^2] \le A_3(1+E[X_0|^2])\cdot e^{A_3t},$$

which proves that $X \in N[0,T]$.

It remains to show that X_t satisfies (5.17). For all n we have

$$(5.27) \qquad Y_t^{(n+1)} = X_0 + \int_0^t b(s,Y_s^{(n)})ds + \int_0^t \sigma(s,Y_s^{(n)})dB_s$$

Now $Y_t^{(n+1)} \to X_t$, $b(s,Y_s^{(n)}) \to b(s,X_s)$
and $\sigma(s,Y_s^{(n)}) \to \sigma(s,X_s)$ as $n\to\infty$,
uniformly in $t \in [0,T]$ for a.s.ω. By (5.26) and the Ito isometry it follows that

$$\int_0^t \sigma(s,Y_s^{(n)})dB_s \to \int_0^t \sigma(s,X_s)dB_s$$

in $L^2(\Omega)$. Therefore, taking the limit of (5.27) as $n \to \infty$ we obtain (5.17) for X_t.

REMARKS.

1) The solution X_t found above is called a strong solution, because the version B_t of Brownian motion is given in advance and the solution X_t constructed from it is $\mathcal{F}_t$-adapted. If we are only given the functions $b(t,x)$ and $\sigma(t,x)$ and ask for a pair of processes $((X_t,\tilde{B}_t),\mathcal{H}_t)$ on a probability space $(\Omega,\mathcal{H},P)$ such that (5.17) holds, then the solution X_t (or more precisely $(X_t,\tilde{B}_t)$) is called a weak solution. Here $\mathcal{H}_t$ is an increasing family of σ-algebras such that X_t is $\mathcal{H}_t$-adapted and $\tilde{B}_t$ is an $\mathcal{H}_t$-Brownian motion, i.e. $\tilde{B}_t$ is a Brownian motion, and $\tilde{B}_t$ is a martingale wrt. $\mathcal{H}_t$ (and so $E[\tilde{B}_{t+h} - \tilde{B}_t | \mathcal{H}_t] = 0$ for all $t,h > 0$). Recall from Chapter III that this allows us to define the Ito integral on the right hand side of (5.17) exactly as before, even though X_t need not be $\mathcal{F}_t$-adapted.

A strong solution is of course also a weak solution, but the converse is not true in general. See the example of Tanaka, explained in Rogers and Williams [1, Sec.V.16].

2) The uniqueness (5.20) that we obtain above is called strong or pathwise uniqueness, while weak uniqueness simply means that any two solutions (weak or strong) are identical in law, i.e. have the same finite-dimensional distributions. See Stroock and Varadhan [1] for results about existence and uniqueness of weak solutions. A general discussion about strong and weak solutions can be found in Krylov and Zvonkin [1].

3) If b and σ satisfy the conditions of Theorem 5.5 then we have

(5.28) A solution (weak or strong) of (5.17) is weakly unique.

To see this we first note that if X_t denotes the strong $\mathcal{F}_t$-adapted solution for the given Brownian motion B_t and $\hat{X}_t$ is an $\mathcal{H}_t$-adapted solution where $\mathcal{F}_t \subset \mathcal{H}_t$ and B_t is an $\mathcal{H}_t$-Brownian motion, then the same uniqueness argument as above applies to show that $X_t = \hat{X}_t$ for all t, a.s. Next, if $\tilde{B}_t$ is another Brownian motion and $\tilde{X}_t$ is constructed from $\tilde{B}_t$ as above, then X_t and $\tilde{X}_t$ must be identical in law. We see this by proving by induction that for each k the two pairs of processes

$$(Y_t^{(k)}, B_t) \quad \text{and} \quad (\tilde{Y}_t^{(k)}, \tilde{B}_t)$$

have the same law. Here $Y_t^{(k)}$, $\tilde{Y}_t^{(k)}$ are the processes defined by (5.24), with Brownian motions B_t, $\tilde{B}_t$, respectively.

This observation will be useful for us in Chapter VII and later, where we will investigate further the properties of processes which are solutions of stochastic differential equations (Ito diffusions).

VI. The Filtering Problem

Problem 3 in the introduction is a special case of the following general filtering problem:

Suppose the state $X_t \in \mathbb{R}^n$ at time t of a system is given by a stochastic differential equation

$$\text{(6.1)} \qquad \frac{dX_t}{dt} = b(t,X_t) + \sigma(t,X_t)\, W_t,$$

where $b\colon \mathbb{R}^{n+1} \to \mathbb{R}^n$, $\sigma\colon \mathbb{R}^{n+1} \to \mathbb{R}^{n\times p}$ satisfy conditions (5.14), (5.15) and W_t is p-dimensional white noise. As discussed earlier the Ito interpretation of this equation is

$$\text{(6.2) (SYSTEM)} \qquad \underline{dX_t = b(t,X_t)dt + \sigma(t,X_t)dU_t},$$

where U_t is p-dimensional Brownian motion. We also assume that the distribution of X_0 is known. (Similarly to the 1-dimensional situation (3.20) there is an explicit several-dimensional formula (in terms of Ito integrals) of the Stratonovich interpretation of (6.1):

$$\text{(6.3)} \qquad dX_t = \tilde{f}(t,X_t)dt + c(t,X_t)dU_t, \quad \text{where}$$

$$\tilde{f}_i(t,x) = f_i(t,x) + \frac{1}{2}\sum_{j=1}^{p}\sum_{k=1}^{n} \frac{\partial\sigma_{ij}}{\partial x_k}\sigma_{kj}\ ;\ 1 \leqslant i \leqslant n).$$

(See Stratonovich [1]).

In the continuous version of the filtering problem we assume that the observations $H_t \in \mathbb{R}^m$ are performed continuously and are of the form

$$\text{(6.4)} \qquad H_t = c(t,X_t) + \gamma(t,X_t) \cdot \tilde{W}_t^{\bullet},$$

where $c : \mathbb{R}^{n+1} \to \mathbb{R}^m$, $\gamma : \mathbb{R}^{n+1} \to \mathbb{R}^{m\times r}$ are functions satisfying (5.14) and $\tilde{W}_t$ denotes r-dimensional white noise, independent of U_t and X_0.

To obtain a tractable mathematical interpretation of (6.4) we introduce

$$\text{(6.5)} \qquad Z_t = \int_0^t H_s ds$$

and thereby we obtain the stochastic integral representation

$$\text{(6.6) (OBSERVATIONS)} \qquad \underline{dZ_t = c(t,X_t)dt + \gamma(t,X_t)dV_t,\ Z_0 = 0}$$

where V_t is r-dimensional Brownian motion, independent of U_t and X_0.

Note that if H_s is known for $0 \leq s \leq t$, then Z_s is also known for $0 \leq s \leq t$ and vica versa. So no information is lost or gained by considering Z_t as our "observations" instead of H_t. But this allows us to obtain a well-defined mathematical model of the situation.

The filtering problem is the following:
Given the observations Z_s satisfying (6.6) for $0 \leq s \leq t$, what is the best estimate $\hat{X}_t$ of the state X_t of the system (6.2) based on these observations?

As we have pointed out earlier, it is necessary to find a precise mathematical formulation of this problem:
By saying that the estimate $\hat{X}_t$ is based on the observations $\{Z_s;\ s \leq t\}$ we mean that

(6.7) $\hat{X}_t(\cdot)$ is $\mathcal{G}_t$-measurable,
where $\mathcal{G}_t$ is the σ-algebra generated by $\{Z_s(\cdot),\ s \leq t\}$.

By saying that $\hat{X}_t$ is the best such estimate we mean that

(6.8) $$\int_\Omega |X_t-\hat{X}_t|^2 dP = E[|X_t-\hat{X}_t|^2] = \inf\{E[|X_t-S|^2];\ S \in \mathcal{K}\}$$

Here - and in the rest of this chapter - $(\Omega,\mathcal{F},P)$ is the probability space corresponding to the $(p+r)$ - dimensional Brownian motion starting at 0, E denotes expectation wrt. P and

(6.9) $$\mathcal{K} = \mathcal{K}_t = \mathcal{K}(Z,t) = \{S:\Omega \to \mathbb{R}^n;\ S \in L^2(P) \text{ and } S \text{ is } \mathcal{G}_t\text{-measurable}\},$$

where $L^2(P) = L^2(\Omega,P)$.

Having found the mathematical formulation of our problem, we now start to study the properties of the solution $\hat{X}_t$.

We first establish the following useful connection between conditional expectation and projection:

LEMMA 6.1. Let $\mathcal{H} \subset \mathcal{F}$ be a σ-algebra and let $X \in L^2(P)$ be $\mathcal{F}$-measurable. Put $\mathcal{N} = \{S \in L^2(P)\ ;\ S$ is $\mathcal{H}$-measurable$\}$ and let $\mathcal{P}_{\mathcal{N}}$ denote the (orthogonal) projection from the Hilbert space $L^2(P)$ into the subspace $\mathcal{N}$. Then

$$\mathcal{P}_{\mathcal{N}}(X) = E[X|\mathcal{H}].$$

Proof. Recall (see Appendix B) that $E[X|\mathcal{H}]$ is by definition the P-unique function from Ω to $\mathbb{R}$ such that

(i) $E[X|\mathcal{H}]$ is $\mathcal{H}$-measurable

(ii) $\int_A E[X|\mathcal{H}]dP = \int_A XdP$ for all $A \in \mathcal{H}$.

Now $\mathcal{P}_{\mathcal{N}}(X)$ is $\mathcal{H}$-measurable and

$$\int_\Omega S(X-\mathcal{P}_{\mathcal{N}}(X))dP = 0 \quad \text{for all} \quad S \in \mathcal{N}.$$

In particular,

$$\int_A (X-\mathcal{P}_{\mathcal{N}}(X))dP = 0 \quad \text{for all} \quad A \in \mathcal{H}$$

i.e.

$$\int_A \mathcal{P}_{\mathcal{N}}(X)dP = \int_A XdP \quad \text{for all} \quad A \in \mathcal{H}.$$

Hence, by uniqueness, $\mathcal{P}_{\mathcal{N}}(X) = E[X|\mathcal{H}]$.

From the general theory of Hilbert spaces we know that the solution $\hat{X}_t$ of the problem (6.8) is given by the projection $\mathcal{P}_{\mathcal{K}_t}(X_t)$. Therefore Lemma 6.1 leads to the following useful result:

THEOREM 6.2.

$$\hat{X}_t = \mathcal{P}_{\mathcal{K}_t}(X_t) = E[X_t | \mathcal{G}_t]$$

This is the basis for the general Fujisaki-Kallianpur-Kunita equation of filtering theory. See for example Davis [2], Kallianpur [1] or Kunita [1].

Here we will concentrate on the linear case, which allows an explicit solution in terms of a stochastic differential equation for $\hat{X}_t$ (the Kalman-Bucy filter):

In the linear filtering problem the system and observation equations have the form:

(6.10) (LINEAR SYSTEM) $dX_t=F(t)X_t dt+C(t)dU_t; F(t)\in \mathbb{R}^{n\times n}, C(t)\in \mathbb{R}^{n\times p}$

(6.11) (LINEAR OBSERVATIONS) $dZ_t=G(t)X_t dt+D(t)dV_t; G(t)\in \mathbb{R}^{m\times n}$,

$D(t)\in \mathbb{R}^{m\times r}$.

To be able to focus on the main ideas in the solution of the filtering problem, we will consider only the 1-dimensional case:

(6.12) (SYSTEM) $dX_t=F(t)X_t+C(t)dU_t$; $F(t)$, $C(t) \in \mathbb{R}$

(6.13) (OBSERVATIONS) $dZ_t=G(t)X_t dt+D(t)dV_t$; $G(t)$, $D(t) \in \mathbb{R}$.

We assume (see (5.14)) that F, G, C, D are bounded on bounded intervals. Based on our interpretation (6.5) of Z_t we assume $Z_0=0$.

We also assume that X_0 is normally distributed and independent of $\{U_t\}$, $\{V_t\}$ and that $E[X_o] = 0$. Finally we assume that $D(t)$ is bounded away from 0 on bounded intervals.

The (important) extension to the several-dimensional case (6.10), (6.11) is technical, but does not require any essentially new ideas. Therefore we shall only state the result for this case after we have discussed the 1-dimensional situation. The reader is encouraged to work out the necessary modifications for the general case for himself or consult Kallianpur [1] or Davis [1] for a full treatment.

From now on we let X_t, Z_t be processes satisfying (6.12), (6.13). Here is an outline of the solution of the filtering problem in this case.

Step 1. Let $\mathscr{L}=\mathscr{L}(Z,t)$ be the closure in $L^2(\Omega)$ of functions which are linear combinations of the form

$$c_1 Z_{s_1}(\omega) + \dots + c_k Z_{s_k}(\omega), \quad \text{with } s_j \leq t.$$

Let

$$\mathscr{P}_{\mathscr{L}} \text{ denote the projection from } L^2(\Omega) \text{ onto } \mathscr{L}.$$

Then, with $\mathscr{K}$ as in (6.9),

$$\hat{X}_t = \mathscr{P}_{\mathscr{K}}(X_t) = E[X_t | \mathscr{G}_t] = \mathscr{P}_{\mathscr{L}}(X_t).$$

Thus, the best Z-measurable estimate of X_t coincides with the best Z-linear estimate of X_t.

Step 2. Replace Z_t by the innovation process N_t:

$$N_t = Z_t - \int_0^t \widehat{(GX)}_s ds, \quad \text{where} \quad \widehat{(GX)}_s = \mathscr{P}_{\mathscr{L}(Z,s)}(G(s)X_s).$$

Then (i) N_t has orthogonal increments, i.e. $E[(N_{t_1}-N_{s_1})(N_{t_2}-N_{s_2})] = 0$ for non-overlapping intervals $[s_1,t_1]$, $[s_2,t_2]$.

(ii) $\mathscr{L}(N,t) = \mathscr{L}(Z,t)$, so $\hat{X}_t = \mathscr{P}_{\mathscr{L}(N,t)}(X_t)$.

Step 3. If we put

$$dR_t = [D^2(t)]^{-\frac{1}{2}} \cdot dN_t,$$

then R_t is a 1-dimensional Brownian motion. Moreover,

$$\mathcal{L}(N,t) = \mathcal{L}(R,t) \quad \text{and}$$

$$\hat{X}_t = \mathcal{P}_{\mathcal{L}(N,t)}(X_t) = \mathcal{P}_{\mathcal{L}(R,t)}(X_t) = \int_0^t \frac{\partial}{\partial s} E(X_t R_s) \cdot dR_s .$$

Step 4. Find expression for X_t by solving the (linear) stochastic differential equation

$$dX_t = F(t)X_t dt + C(t)dU_t .$$

Step 5. Substitute the formula for X_t from Step 4 into $E[X_t R_s]$ and use Step 3 to obtain a stochastic differential equation for $\hat{X}_t$:

$$d\hat{X}_t = \frac{\partial}{\partial s} E[X_t R_s]_{s=t} dR_t + \left(\int_0^t \frac{\partial^2}{\partial t \partial s} E[X_t R_s] dR_s\right) dt \quad \text{etc.}$$

Before we proceed to establish Steps 1-5, let us consider a simple, but motivating example:

EXAMPLE 6.3. Suppose $X, W_1, W_2, \ldots$ are independent real random variables, $E[X] = E[W_j] = 0$ for all j, $E[X^2] = a^2$, $E[W_j^2] = m^2$ for all j. Put $Z_j = X + W_j$.

What is the best linear estimate $\hat{X}$ of X based on $\{Z_j ; j \leq k\}$? More precisely, let

$$\mathcal{L} = \mathcal{L}(Z,k) = \{c_1 Z_1 + \ldots + c_k Z_k \; ; \; c_1, \ldots, c_k \in R\}.$$

Then we want to find

$$\hat{X}_k = \mathcal{P}_k(X),$$

where $\mathcal{P}_k$ denotes the projection onto $\mathcal{L}(Z,k)$.

We use the Gram-Schmidt procedure to obtain random variables $A_1, A_2, \ldots$ such that

(i) $E[A_i A_j] = 0$ for $i \neq j$

(ii) $\mathcal{L}(A,k) = \mathcal{L}(Z,k)$ for all k.

Then

$$\hat{X}_k = \sum_{j=1}^{k} \frac{E[XA_j]}{E[A_j^2]} \cdot A_j \quad \text{for} \quad k = 1, 2, \ldots . \tag{6.14}$$

We obtain a recursive relation between $\hat{X}_k$ and $\hat{X}_{k-1}$ from this by observing that

(6.15) $A_j = Z_j - \hat{X}_{j-1}$,
which follows from

$$A_j = Z_j - \mathscr{P}_{j-1}(Z_j) = Z_j - \mathscr{P}_{j-1}(X), \quad \text{since } \mathscr{P}_{j-1}(W_j) = 0.$$

By (6.15)

$$E[XA_j] = E[X(Z_j-\hat{X}_{j-1})] = E[X(X-\hat{X}_{j-1})] = E[(X-\hat{X}_{j-1})^2]$$

and

$$E[A_j^2] = E[(X+W_j-\hat{X}_{j-1})^2] = E[(X-\hat{X}_{j-1})^2] + m^2.$$

Hence

$$\hat{X}_k = \hat{X}_{k-1} + \frac{E[(X-\hat{X}_{k-1})^2]}{E[(X-\hat{X}_{k-1})^2]+m^2} \cdot (Z_k - \hat{X}_{k-1}). \tag{6.16}$$

If we introduce

$$\bar{Z}_k = \frac{1}{k} \sum_{j=1}^{k} Z_j,$$

then this can be simplified to

$$\hat{X}_k = \frac{a^2}{a^2 + \frac{1}{k} \cdot m^2} \cdot \bar{Z}_k. \tag{6.17}$$

(This can be seen as follows:
Put

$$\alpha_k = \frac{a^2}{a^2 + \frac{1}{k} m^2}, \quad U_k = \alpha_k \bar{Z}_k.$$

Then

(i) $U_k \in \mathscr{L}(Z,k)$

(ii) $X - U_k \perp \mathscr{L}(Z,k)$, since

$$\begin{aligned} E[(X-U_k)Z_i] &= E[XZ_i] - \alpha_k E[\bar{Z}_k Z_i] \\ &= E[X(X+W_i)] - \alpha_k \cdot \frac{1}{k} \cdot \sum_j E[Z_j Z_i] \\ &= a^2 - \frac{1}{k} \cdot \alpha_k \cdot \sum_j E[(X+W_j)(X+W_i)] \\ &= a^2 - \frac{1}{k} \cdot \alpha_k [k \cdot a^2 + m^2] = 0 \ . \ ) \end{aligned}$$

The result can be interpretedd as follows:
For large k we put $\hat{X}_k \approx \bar{Z}_k$, while for small k the relation between a^2 and m^2 becomes more important. If $m^2 \gg a^2$, the observations are to a large extent neglected (for small k) and $\hat{X}_k$ is put equal to its mean value, 0.

This example gives the motivation for our approach:
We replace the process Z_t by an orthogonal increment process N_t (Step 2) in order to obtain a representation for $\hat{X}_t$ analogous to (6.14). Such a representation is obtained in Step 3, after we have identified the best linear estimate with the best measurable estimate (Step 1) and established the connection between N_t and Brownian motion.

STEP 1. Z-LINEAR AND Z-MEASURABLE ESTIMATES

LEMMA 6.4. Let $X, Z_s; s \leq t$ be random variables in $L^2(P)$ and assume that

$$(X, Z_{s_1}, Z_{s_2}, \ldots, Z_{s_n}) \in \mathbb{R}^{n+1}$$

has a normal distribution with mean 0, for all $s_1, s_2, \ldots, s_n \leq t$, $n \geq 1$. Then

$$\mathcal{P}_{\mathcal{L}}(X) = E[X|\mathcal{G}] = \mathcal{P}_{\mathcal{K}}(X).$$

In other words, the best Z-linear estimate for X coincides with the best Z-measurable estimate in this case.

Proof. Put $\check{X} = \mathcal{P}_{\mathcal{L}}(X)$, $\tilde{X} = X - \check{X}$.
Then we claim that $\tilde{X}$ is independent of $\mathcal{G}$:
Recall that a random variable $(Y_1, \ldots, Y_k) \in \mathbb{R}^k$ is normal iff $c_1Y_1 + \cdots + c_kY_k$ is normal, for all choices of $c_1, \ldots, c_k \in \mathbb{R}$. And an L^2-limit of normal variables is again normal (Appendix A). Therefore

$$(\tilde{X}, Z_{s_1}, \ldots, Z_{s_n}) \text{ is normal for all } s_1, \ldots, s_n \leq t.$$

Since $E[\tilde{X}Z_{s_j}] = 0$, $\tilde{X}$ and Z_{s_j} are uncorrelated, for $1 \leq j \leq n$. It follows (Appendix A) that

$$\tilde{X} \text{ and } (Z_{s_1}, \ldots, Z_{s_n}) \text{ are independent.}$$

So $\tilde{X}$ is independent from $\mathcal{G}$ as claimed. But then

$$E[\chi_G(X-\check{X})] = E[\chi_G\tilde{X}] = E[\chi_G]\cdot E[\tilde{X}] = 0 \quad \text{for all } G\in\mathcal{G}$$

i.e. $\int_G XdP = \int_G \check{X}dP$. Since $\check{X}$ is $\mathcal{G}$-measurable, we conclude that $\check{X} = E[X|\mathcal{G}]$.

There is a curious interpretation of this result: Suppose $X,\{Z_t\}_{t\in T}$ are $L^2(P)$-functions with given covariances. Among all possible distributions of

$$(X,Z_{t_1},\dots,Z_{t_n})$$

with these covariances, the normal distributions will be the "worst" wrt. estimation, in the following sense: For any distribution we have

$$E[(X-E[X|\mathcal{G}])^2] \leq E[(X-\mathcal{P}_{\mathcal{L}}(X))^2],$$

with equality for the normal distribution, by Lemma 6.4. (Note that the quantity on the right hand side only depends on the covariances, not on the distribution we might choose to obtain these covariances). For a broad discussion of similar conclusions, based on an information theoretical game between nature and the observer, see Topsöe [1].

Finally, to be able to apply Lemma 6.4 to our filtering problem, we need the following result:

LEMMA 6.5.

$$M_t=\begin{bmatrix} X_t \\ Z_t \end{bmatrix} \in \mathbb{R}^2 \quad \text{is a Gaussian process.}$$

Proof. We may regard M_t as the solution of a 2-dimensional linear stochastic differential equation of the form

$$(6.18)\qquad dM_t = H(t)M_t dt + K(t)dB_t,\quad M_0=\begin{bmatrix} X_0 \\ 0 \end{bmatrix};$$

where $H(t)\in\mathbb{R}^{2\times 2}$, $K(t)\in\mathbb{R}^{2\times 2}$ and B_t is 2-dimensional Brownian motion. As in Example 5.3 in Chapter V we may solve equation (6.18) explicitly:

$$(6.19)\qquad M_t= \exp[\int_0^t H(s)ds]\{M_0+\int_0^t \exp(-\int_0^s H(r)dr)K(s)dB_s\}$$

and from this expression it follows that $\{M_t\}$ is a Gaussian process.

STEP 2. THE INNOVATION PROCESS

Before we introduce the innovation process we will establish a useful representation of the functions in the space

$$\mathscr{L}(Z,T) = \text{the closure in } L^2(\Omega) \text{ of all linear combinations}$$
$$c_1 Z_{t_1} + \dots + c_k Z_{t_k} \; ; \; 0 \le t_i \le T.$$

If $f \in L^2[0,T]$, note that

$$E[(\int_0^T f(t)dZ_t)^2] = E[(\int_0^T f(t)G(t)X_t dt)^2] + E[(\int_0^T f(t)D(t)dV_t)^2]$$
$$+ 2E[(\int_0^T f(t)G(t)X_t dt)\cdot(\int_0^T f(t)D(t)dV_t)].$$

Since

$E[(\int_0^T f(t)G(t)X_t dt)^2] \le A_1 \cdot \int_0^T f(t)^2 dt$ by the Cauchy-Schwartz inequality,

$E[(\int_0^T f(t)D(t)dV_t)^2] = \int_0^T f(t)^2 D^2(t)dt$ by the Ito isometry

and

$\{X_t\}$, $\{V_t\}$ are independent, we conclude that

$$A_0 \cdot \int_0^T f^2(t)dt \le E[(\int_0^T f(t)dZ_t)^2] \le A_1 \cdot \int_0^T f^2(t)dt, \tag{6.20}$$

for some constants A_0, A_1 not depending on f.

We can now show

LEMMA 6.6. $\mathscr{L}(Z,T) = \{\int_0^T f(t)dZ_t \; ; \; f \in L^2[0,T]\}$.

Proof. Denote the right hand side by $\mathscr{N}(Z,T)$. It is enough to show that

a) $\mathscr{N}(Z,T) \subset \mathscr{L}(Z,T)$

b) $\mathscr{N}(Z,T)$ contains all linear combinations of the form
$$c_1 Z_{t_1} + \dots + c_k Z_{t_k} \; ; \; 0 \le t_i \le T$$

c) $\mathscr{N}(Z,T)$ is closed in $L^2(P)$

a): This follows from the fact that

$$\int_0^T f(t)dZ_t = \lim_{n\to\infty} \sum_j f(j\cdot 2^{-n})\cdot (Z_{(j+1)2^{-n}} - Z_{j\cdot 2^{-n}}).$$

b): Suppose $0 \leq t_1 < t_2 < \ldots < t_k \leq T$. We can write

$$\sum_{i=1}^{k} c_i Z_{t_i} = \sum_{j=0}^{k-1} c_j' \Delta Z_j = \sum_{j=0}^{k-1} \int_{t_j}^{t_{j+1}} c_j' \, dZ_t = \int_0^T \Big(\sum_{j=0}^{k-1} c_j' \chi_{[t_j, t_{j+1})}(t)\Big) dZ_t .$$

c): This follows from (6.20) and the fact that $L^2[0,T]$ is complete.

Now we define the <u>innovation process</u> N_t as follows:

(6.21) $$N_t = Z_t - \int_0^t (\widehat{GX})_s ds, \quad \text{where} \quad (\widehat{GX})_s = \mathcal{P}_{\mathcal{L}(Z,s)}(G(s)X_s).$$

(6.22) i.e. $$dN_t = (G(t)X_t - (\widehat{GX})_t)dt + D(t)dV_t .$$

<u>LEMMA 6.7</u>.

(i) N_t has orthogonal increments

(ii) $E[N_t^2] = \int_0^t D^2(s)ds$

(iii) $\mathcal{L}(N,t) = \mathcal{L}(Z,t)$ for all $t \geq 0$

(iv) N_t is a Gaussian process

<u>Proof</u>.

(i): If $s < t$ and $Y \in \mathcal{L}(Z,s)$ we have

$$E[(N_t - N_s)Y] = E[(\int_s^t (G(r)X_r - (\widehat{GX})_r)dr + \int_s^t D(r)dV_r)Y]$$

$$= \int_s^t E[(GX-(GX)) \cdot Y]dr + E[(\int_s^t DdV)\cdot Y] = 0,$$

since $G(r)X_r - (\widehat{GX})_r \perp \mathcal{L}(Z,r) \supset \mathcal{L}(Z,s)$ for $r \geq s$ and V has independent increments.

(ii): By Ito's formula, with $g(t,x) = x^2$, we have

$$d(N_t^2) = 2N_t dN_t + \tfrac{1}{2} \cdot 2(dN_t)^2 = 2N_t dN_t + D^2 dt.$$

So

$$E[N_t^2] = E[\int_0^t 2N_s dN_s] + \int_0^t D^2(s)ds.$$

Now

$$\int_0^t N_s dN_s = \lim_{\Delta t_j \to 0} \sum N_{t_j}[N_{t_{j+1}} - N_{t_j}],$$

so since N has orthogonal increments we have

$E[\int_0^t N_s dN_s] = 0$, and (ii) follows.

(iii): It is clear that $\mathcal{L}(N,t) \subset \mathcal{L}(Z,t)$ for all $t>0$. To establish the opposite inclusion we use Lemma 6.6 So choose $f \in L^2[0,t]$ and let us see what functions can be obtained in the form

$$\int_0^t f(s)dN_s = \int_0^t f(s)dZ_s - \int_0^t f(r)(GX)_r dr$$

$$= \int_0^t f(s)dZ_s - \int_0^t f(r)[\int_0^r g(r,s)dZ_s]dr$$

$$= \int_0^t [f(s) - \int_s^t f(r)g(r,s)dr]dZ_s,$$

where we have used Lemma 6.5 to write

$$(\widehat{GX})_r = \int_0^r g(r,s)dZ_s \quad \text{for some} \quad g(r,\cdot) \in L^2[0,r].$$

From the theory of Volterra integral equations (see e.g. Davis [1], p. 125) there exists for all $h \in L^2[0,t]$ an $f \in L^2[0,t]$ such that

$$f(s) - \int_s^t f(r)g(r,s)dr = h(s).$$

So by choosing $h = \chi_{[0,t_1]}$ where $0 \le t_1 \le t$, we obtain

$$\int_0^t f(s)dN_s = \int_0^t \chi_{[0,t_1]}(s)dZ_s = Z_{t_1},$$

which shows that $\mathcal{L}(N,t) \supset \mathcal{L}(Z,t)$.

(iv): $(\widehat{GX})_t$ is a limit (in $L^2(P)$) of linear combinations of the form

$M = c_1 Z_{s_1} + \dots + c_k Z_{s_k}$, where $s_k \le t$.

Therefore

$((\widehat{GX})_{t_1}, \dots, (\widehat{GX})_{t_m})$

is a limit of m-dimensional random variables $(M^{(1)},\dots,M^{(m)})$ whose components $M^{(j)}$ are linear combinations of this form. $(M^{(1)},\dots,M^{(m)})$ has a normal distribution since $\{Z_t\}$ is Gaussian, and therefore the limit has. Hence $(\widehat{GX})_t$ is Gaussian. It follows that

$$N_t = Z_t - \int_0^t (\widehat{GX})_s ds$$

is Gaussian, by a similar argument.

STEP. 3. THE INNOVATION PROCESS AND BROWNIAN MOTION

Let $N_t = Z_t - \int_0^t (\widehat{GX})_s ds$ be the innovation process defined in Step 2. Recall that we have assumed that $D(t)$ is bounded away from 0 on bounded intervals. Define the process $R_t(\omega)$ by

$$dR_t = \frac{1}{D(t)} \cdot dN_t(\omega) \; ; \; t \geq 0. \tag{6.23}$$

LEMMA 6.8. R_t is a 1-dimensional Brownian motion.

Proof. Observe that

(i) R_t has continuous paths

(ii) R_t has orthogonal increments (since N_t has)

(iii) R_t is Gaussian (since N_t is)

(iv) $E[R_t] = 0$ and $E[R_t R_s] = \min(s,t)$

To prove the last assertion in (iv), note that by Ito's formula

$$d(R_t^2) = 2R_t dR_t + (dR_t)^2 = 2R_t dR_t + dt,$$

so, since R_t has orthogonal increments,

$$E[R_t^2] = E[\int_0^t ds] = t.$$

Therefore, if $s < t$,

$$E[R_t R_s] = E[(R_t - R_s)R_s] + E[R_s^2] = E[R_s^2] = s.$$

Properties (i), (iii) and (iv) constitute one of the many characterizations of a 1-dimensional Brownian motion (see Williams [1], I. 15). (Alternatively, we could easily deduce that R_t has stationary, independent increments and therefore - by continuity - must be Brownian motion, by the result previously referred to in the beginning of Chapter III). For a general characterization of Brownian motion see Corollary 8.13.

Since

$$\mathcal{L}(N,t) = \mathcal{L}(R,t)$$

we conclude that

$$\hat{X}_t = \mathcal{P}_{\mathcal{L}(R,t)}(X_t)$$

It turns out that the projection down to the space $\mathcal{L}(R,t)$ can be described very nicely: (compare with formula (6.14) in Example 6.3!)

LEMMA 6.9.

$$(6.24) \qquad \hat{X}_t = \int_0^t \frac{\partial}{\partial s} E[X_t R_s] \cdot dR_s .$$

Proof. From Lemma 6.6 we know that

$$\hat{X}_t = \int_0^t g(s) dR_s \qquad \text{for some} \quad g \in L^2[0,t].$$

We have

$$(X_t - \hat{X}_t) \perp \int_0^t f(s) dR_s \qquad \text{for all} \quad f \in L^2[0,t].$$

Therefore

$$E[X_t \int_0^t f(s)dR_s] = E[\hat{X}_t \int_0^t f(s)dR_s] = E[\int_0^t g(s)dR_s \int_0^t f(s)dR_s]$$

$$= E[\int_0^t g(s)f(s)ds] = \int_0^t g(s)f(s)ds, \quad \text{for all} \quad f \in L^2[0,1]$$

where we have used the Ito isometry. In particular, if we choose

$$f = \chi_{[0,r]} \quad \text{for some} \quad r \leq t,$$

we obtain

$$E[X_t R_r] = \int_0^r g(s)ds$$

or

$$g(r) = \frac{\partial}{\partial r} E[X_t R_r], \quad \text{as asserted.}$$

This completes Step 3.

STEP 4. AN EXPLICIT FORMULA FOR X_t

This is easily obtained using Ito's formula, as in the examples in Chapter V. the result is

$$X_t = \exp(\int_0^t F(s)ds)[X_0 + \int_0^t \exp(-\int_0^s F(u)du)C(s)dU_s]$$

$$= \exp\left(\int_0^t F(s)ds\right)X_0 + \int_0^t \exp\left(\int_s^t F(u)du\right)C(s)dU_s .$$

Or, more generally if $0 \leq r \leq t$

$$X_t = \exp\left(\int_r^t F(s)ds\right)X_r + \int_r^t \exp\left(\int_s^t F(u)du\right)C(s)dU_s . \tag{6.25}$$

STEP 5. THE STOCHASTIC DIFFERENTIAL EQUATION FOR $\hat{X}_t$

We now combine the previous steps to obtain the solution of the filtering problem, i.e. a stochastic differential equation for $\hat{X}_t$. Starting with the formula from Lemma 6.9 (6.24)

$$\hat{X}_t = \int_0^t f(s,t)dR_s ,$$

where

$$f(s,t) = \frac{\partial}{\partial s} E[X_t R_s] , \tag{6.26}$$

we use that

$$R_s = \int_0^s \frac{G(r)}{D(r)} \cdot (X_r - \hat{X}_r)dr + V_s \qquad \text{(from (6.22) and (6.23))}$$

and obtain

$$E[X_t R_s] = \int_0^s \frac{G(r)}{D(r)} E[X_t \tilde{X}_r]dr ,$$

where

$$\tilde{X}_r = X_r - \hat{X}_r . \tag{6.27}$$

Using formula (6.25) for X_t, we obtain

$$E[X_t \tilde{X}_r] = \exp\left(\int_r^t F(v)dv\right)E[X_r \tilde{X}_r] = \exp\left(\int_r^t F(v)dv\right) \cdot S(r) ,$$

where

$$S(r) = E[(\tilde{X}_r)^2] , \tag{6.28}$$

i.e. the mean square error of the estimate at time $r \geq 0$. Thus

$$E[X_t R_s] = \int_0^s \frac{G(r)}{D(r)} \cdot \exp\left(\int_r^t F(v)dv\right)S(r)dr$$

so that

$$f(s,t) = \frac{G(s)}{D(s)} \cdot \exp\left(\int_s^t F(v)dv\right)S(s) . \tag{6.29}$$

We claim that $S(t)$ satisfies the (deterministic) differential equation

$$\frac{dS}{dt} = 2F(t)S(t) - \frac{G^2(t)}{D^2(t)} \cdot S^2(t) + C^2(t) \quad \text{(The Riccati equation).} \tag{6.30}$$

To prove (6.30) note that by the Pythagorean theorem, (6.24) and the Ito isometry

$$S(t) = E[(X_t-\hat{X}_t)^2] = E[(X_t)^2] - E[(\hat{X}_t)^2]$$

$$= T(t) - \int_0^t f(s,t)^2 ds, \tag{6.31}$$

where

$$T(t) = E[X_t^2]. \tag{6.32}$$

Now by (6.25) and the Ito isometry we have

$$T(t) = \exp(2\int_0^t F(s)ds)E[X_0^2] + \int_0^t \exp(2\int_0^t F(u)du)C^2(s)ds,$$

using that X_0 is independent of $\{U_s\}_{s\geq 0}$. So

$$\frac{dT}{dt} = 2F(t)\cdot\exp(2\int_0^t F(s)ds)E[X_0^2] + C^2(t) + \int_0^t 2F(t)\exp(2\int_0^t F(u)du)C^2(s)ds$$

i.e.

$$\frac{dT}{dt} = 2F(t)T(t) + C^2(t). \tag{6.33}$$

Substituting in (6.31) we obtain

$$\frac{dS}{dt} = \frac{dT}{dt} - f(t,t)^2 - \int_0^t 2\cdot f(s,t) \cdot \frac{\partial}{\partial t} f(s,t)ds$$

$$= 2F(t)T(t) + C^2(t) - \frac{G^2(t)S^2(t)}{D^2(t)} - 2\int_0^t f^2(s,t)F(t)ds$$

$$= 2F(t)S(t) + C^2(t) - \frac{G^2(t)S^2(t)}{D^2(t)}, \quad \text{which is (6.30).}$$

We are now ready for the stochastic differential equation for $\hat{X}_t$:

From the formula

$$\hat{X}_t = \int_0^t f(s,t)dR_s$$

it follows that

(6.34) $$d\hat{X}_t = f(t,t)dR_t + \left(\int_0^t \frac{\partial}{\partial t} f(s,t)dR_s\right)dt,$$

since

$$\int_0^u \left(\int_0^t \frac{\partial}{\partial t} f(s,t)dR_s\right)dt = \int_0^u \left(\int_s^u \frac{\partial}{\partial t} f(s,t)dt\right)dR_s$$

$$= \int_0^u (f(s,u)-f(s,s))dR_s = \hat{X}_u - \int_0^u f(s,s)dR_s.$$

So

$$d\hat{X}_t = \frac{G(t)S(t)}{D(t)} dR_t + \left(\int_0^t f(s,t)dR_s\right)F(t)dt$$

or

(6.35) $$d\hat{X}_t = F(t) \cdot \hat{X}_t dt + \frac{G(t)S(t)}{D(t)} dR_t.$$

If we substitute

$$dR_t = \frac{1}{D(t)} \cdot [dZ_t - G(t)\hat{X}_t dt]$$

we obtain

(6.36) $$d\hat{X}_t = \left(F(t) - \frac{G^2(t)S(t)}{D^2(t)}\right)\hat{X}_t dt + \frac{G(t)S(t)}{D^2(t)} dZ_t.$$

So the conclusion is:

THEOREM 6.10. (THE KALMAN - BUCY FILTER)
The solution $\hat{X}_t = E[X_t|\mathcal{G}_t]$ of the linear filtering problem (6.12), (6.13) satisfies the stochastic differential equation

$$d\hat{X}_t = F(t)\hat{X}_t dt + \frac{G(t)S(t)}{D(t)} dR_t; \quad \hat{X}_0 = E[X_0]$$

or

$$d\hat{X}_t = \left(F(t) - \frac{G^2(t)S(t)}{D^2(t)}\right)\hat{X}_t dt + \frac{G(t)S(t)}{D^2(t)} dZ_t; \quad \hat{X}_0 = E[X_0]$$

where

$S(t) = E[(X_t-\hat{X}_t)^2]$ satisfies the (deterministic) Riccati equation

$$\frac{dS}{dt} = 2F(t)S(t) - \frac{G^2(t)}{D^2(t)} S^2(t)+C^2(t), \quad S(0) = E[(X_0-E[X_0])^2].$$

EXAMPLE 6.11. (NOISY OBSERVATIONS OF A CONSTANT PROCESS)
Consider the very simple case

(system) $dX_t = 0$, i.e. $X_t = X_0$, $E[X_0] = 0$, $E[X_0^2] = a^2$

(observations) $dZ_t = X_t dt + m \cdot dV_t$, $Z_0 = 0$

(corresponding to

$$H_t = \frac{dZ}{dt} = X_t + m \cdot W_t, \quad W_t = \text{white noise}).$$

First we solve the corresponding Riccati equation for

$$S(t) = E[(X_t - \hat{X}_t)^2]:$$

$$\frac{dS}{dt} = -\frac{1}{m^2} \cdot S^2, \quad S(0) = a^2$$

i.e.

$$S(t) = \frac{a^2 m^2}{m^2 + a^2 t} \; ; \; t \geq 0.$$

This gives the following equation for $\hat{X}_t$:

$$d\hat{X}_t = -\frac{a^2}{m^2 + a^2 t} \cdot \hat{X}_t dt + \frac{a^2}{m^2 + a^2 t} \cdot dY_t \; ; \; \hat{X}_0 = E[X_0] = 0$$

or

$$d(\hat{X}_t \cdot \exp(\int \frac{a^2}{m^2 + a^2 t} dt)) = \exp(\int \frac{a^2}{m^2 + a^2 t} dt) \cdot \frac{a^2}{m^2 + a^2 t} dY_t$$

which gives

(6.37) $$\hat{X}_t = \frac{a^2}{m^2 + a^2 t} \cdot Y_t \; ; \; t \geq 0.$$

This is the continuous analogue of Example 6.3.

<u>EXAMPLE 6.12</u>. (NOISY OBSERVATIONS OF BROWNIAN MOTION)

If we modify the preceding example slightly, so that

(system) $dX_t = c dU_t$, $E[X_0] = 0$, $E[X_0^2] = a^2$, c constant

and as before

(observations) $dZ_t = X_t dt + m dV_t$,

the Riccati equation becomes

$$\frac{dS}{dt} = -\frac{1}{m^2} \cdot S^2 + c^2, \quad S(0) = a^2$$

or

$$\frac{m^2 dS}{m^2 c^2 - S^2} = dt, \quad (S \neq mc).$$

This gives

$$\left| \frac{mc+S}{mc-S} \right| = K \cdot \exp(\frac{2ct}{m}) \; ; \; K = \left| \frac{mc+a^2}{mc-a^2} \right| .$$

Or

$$S(t) = \begin{cases} mc \cdot \dfrac{K\cdot\exp(\frac{2ct}{m})-1}{K\cdot\exp(\frac{2ct}{m})+1} ; & \text{if } S(0) < mc \\ mc \quad (\text{constant}) & \text{if } S(0) = mc \\ mc \cdot \dfrac{K\cdot\exp(\frac{2ct}{m})+1}{K\cdot\exp(\frac{2ct}{m})-1} & \text{if } S(0) > mc \end{cases}$$

Thus in all cases the mean square error $S(t)$ tends to mc as $t\to\infty$.

For simplicity let us put $a = 0$, $m = c = 1$. Then

$$S(t) = \frac{\exp(2t)-1}{\exp(2t)+1} = \tanh(t).$$

The equation for $\hat{X}_t$ is

$$d\hat{X}_t = -\tanh(t)\cdot\hat{X}_t dt + \tanh(t)\cdot dZ_t, \quad \hat{X}_0 = 0$$

or

$$d(\cosh(t)\cdot\hat{X}_t) = \sinh(t)\cdot dZ_t.$$

So

$$\hat{X}_t = \frac{1}{\cosh(t)}\cdot\int_0^t \sinh(s)dZ_s.$$

If we return to the interpretation of Z_t:

$$Z_t = \int_0^t H_s ds,$$

where H_s are the "original" observations (see (6.3)), we can write

$$\hat{X}_t = \frac{1}{\cosh(t)}\cdot\int_0^t \sinh(s)\cdot H_s ds, \tag{6.38}$$

so $\hat{X}_t$ is approximately (for large t) a weighted average of the observations H_s, with increasing emphasis on observations as time increases.

<u>REMARK</u>. It is interesting to compare formula (6.38) with established formulas in forecasting. For example, the <u>exponentially weighted moving average</u> $\tilde{X}_n$ suggested by C.C. Holt in 1958 is given by

$$\tilde{X}_n = (1-\alpha)^n Z_0 + \alpha\sum_{k=1}^{n}(1-\alpha)^{n-k}Z_k,$$

where α is some constant; $0 \leq \alpha \leq 1$. See The Open University [1, p. 16].

This may be written

$$\tilde{X}_n = \beta^{-n} Z_0 + (\beta-1)\beta^{-n} \sum_{k=1}^{n} \beta^k Z_k ,$$

where $\beta = \frac{1}{1-\alpha}$ (assuming $\alpha < 1$), which is a discrete version of (6.38), or - more precisely - of the formula corresponding to (6.38) in the general case when $a \neq 0$ and m, c not necessarily equal to 1.

EXAMPLE 6.13. (ESTIMATION OF A PARAMETER)

Suppose we want to estimate the value of a (constant) parameter θ, based on observations Z_t satisfying the model

$$dZ_t = \theta \cdot M(t)dt + N(t)dB_t ,$$

where $M(t)$, $N(t)$ are known functions. In this case the stochastic differential equation for θ is of course

$$d\theta = 0,$$

so the Riccati equation for $S(t) = E[(\theta-\hat{\theta}_t)^2]$ is

$$\frac{dS}{dt} = -\left(\frac{M(t)S(t)}{N(t)}\right)^2$$

which gives

$$S(t) = \left(S_0^{-1} + \int_0^t M(s)^2 N(s)^{-2} ds\right)^{-1}$$

and the Kalman-Bucy filter is

$$d\hat{\theta}_t = \frac{M(t)S(t)}{N(t)^2}(dZ_t - M(t)\hat{\theta}_t dt).$$

This can be written

$$\left(S_0^{-1} + \int_0^t M(s)^2 N(s)^{-2} ds\right) d\hat{\theta}_t + M(t)^2 N(t)^{-2} \hat{\theta}_t dt =$$

$$M(t)N(t)^{-2} dZ_t .$$

We recognice the left hand side as

$$d\left(\left(S_0^{-1} + \int_0^t M(s)^2 N(s)^{-2} ds\right)\hat{\theta}_t\right)$$

so we obtain

$$\hat{\theta}_t = \frac{\hat{\theta}_0 S_0^{-1} + \int_0^t M(s)N(s)^{-2} dZ_s}{S_0^{-1} + \int_0^t M(s)^2 N(s)^{-2} ds} .$$

This estimate coincides with the maximum likelihood estimate in classical estimation theory if $S_0^{-1} = 0$. See Lipster and Shiryayev

[2]. For more information about estimates of drift parameters in diffusions and generalizations, see for example Aase [1], Brown and Hewitt [1] and Taraskin [1].

EXAMPLE 6.14. (NOISY OBSERVATIONS OF POPULATION GROWTH)
Consider a simple growth model (r constant)

$$dX_t = r \cdot X_t dt, \quad E[X_0] = b > 0, \quad E[(X_0-b)^2] = a^2,$$

with observations

$$dY_t = X_t dt + m \cdot dV_t \; ; \quad m \text{ constant.}$$

The corresponding Riccati equation

$$\frac{dS}{dt} = 2rS - \frac{1}{m^2} S^2, \quad S(0) = a^2,$$

gives the logistic curve

$$S(t) = \frac{2rm^2}{1+Ke^{-2rt}} \; ; \quad \text{where } K = \frac{2rm^2}{a^2} - 1.$$

So the equation for $\hat{X}_t$ becomes

$$d\hat{X}_t = (r - \frac{S}{m^2})\hat{X}_t dt + \frac{S}{m^2} dZ_t \; ; \quad \hat{X}_0 = E[X_0] = b.$$

For simplicity let us assume that $a^2 = 2rm^2$, so that

$$S(t) = 2rm^2 \quad \text{for all } t.$$

(In the general case $S(t) \to 2rm^2$ as $t \to \infty$, so this is not an unreasonable approximation for large t).
Then we get

$$d(\exp(rt) \cdot \hat{X}_t) = \exp(rt) 2r dZ_t, \quad \hat{X}_0 = b$$

or

$$\hat{X}_t = \exp(-rt)[\int_0^t 2r \cdot \exp(rs) dZ_s + b].$$

As in Example 6.12 this may be written

(6.39) $$\hat{X}_t = \exp(-rt)[\int_0^t 2r \cdot \exp(rs) H_s ds + b], \quad \text{if } Z_t = \int_0^t H_s ds.$$

For example, if $H_s = \beta$ (constant) for $0 \leq s \leq t$, then

$$\hat{X}_t = 2\beta - (2\beta - b)\exp(-rt) \to 2\beta \quad \text{as } t \to \infty.$$

If $H_s = \beta \cdot \exp(\alpha s)$, $s \geq 0$ (α constant), we get

$$\hat{X}_t = \exp(-rt)\left[\frac{2r\beta}{r+\alpha} \cdot (\exp(r+\alpha)t-1) + b\right]$$

$$\approx \frac{2r\beta}{r+\alpha} \cdot \exp \alpha t \quad \text{for large } t.$$

Thus, only if $\alpha=r$, i.e. $H_s = \beta \cdot \exp(rs)$; $s \geq 0$, does the filter "believe" the observations in the long run. And only if $\alpha=r$ and $\beta=b$, i.e. $H_s = b \cdot \exp(rs)$; $s \geq 0$, does the filter "believe" the observations at all times.

EXAMPLE 6.15. (CONSTANT COEFFICIENTS - GENERAL DISCUSSION)
Now consider the system

$$dX_t = F\,X_t dt + C\,dU_t, \quad F, C \text{ constants}, \neq 0$$

with observations

$$dZ_t = G\,X_t dt + D\,dV_t, \quad G, D \text{ constants}, \neq 0.$$

The corresponding Riccati equation

$$S' = 2FS - \frac{G^2}{D^2} \cdot S^2 + C^2, \quad S(0) = a^2$$

has the solution

$$S(t) = \frac{\alpha_1 - K \cdot \alpha_2 \cdot \exp\left(\frac{(\alpha_2-\alpha_1)G^2 t}{D^2}\right)}{1 - K \cdot \exp\left(\frac{(\alpha_2-\alpha_1)G^2 t}{D^2}\right)},$$

where

$$\alpha_1 = G^{-2}(FD^2 - D\sqrt{F^2D^2+G^2C^2})$$

$$\alpha_2 = G^2(FD^2 + D\sqrt{F^2D^2+G^2C^2})$$

and

$$K = \frac{a^2-\alpha_1}{a^2-\alpha_2}$$

This gives the solution for $\hat{X}_t$ of the form

$$\hat{X}_t = \exp\left(\int_0^t H(s)ds\right) \cdot \hat{X}_0 + \frac{G^2}{D^2}\int_0^t \exp\left(\int_s^t H(u)du\right)S(s)dZ_s,$$

where

$$H(s) = F - \frac{G^2}{D^2} \cdot S(s).$$

For a large s we have $S(s) \approx \alpha_2$. This gives

$$\hat{X}_t \approx \hat{X}_0 \cdot \exp((F - \frac{G^2\alpha_2}{D^2})t) + \frac{G\alpha_2}{D^2} \cdot \int_0^t \exp((F \cdot \frac{G^2\alpha_2}{D^2})(t-s)dZ_s$$

$$(6.40) \qquad = \hat{X}_0 \cdot \exp(-\beta t) + \frac{G\alpha_2}{D^2} \cdot \exp(-\beta t) \cdot \int_0^t \exp(\beta s)dZ_s$$

where $\beta = D^{-1}\sqrt{F^2D^2+G^2C^2}$.

So we get approximately the same behaviour as in the previous example.

Finally we formulate the solution of the n-dimensional linear filtering problem (6.10), (6.11):

Regard vectors in $\mathbb{R}^m$ as $m\times 1$ matrices.

Put

$$(6.41) \qquad S(t) = E[(X_t-\hat{X}_t)(X_t-\hat{X}_t)^T] \in \mathbb{R}^{n\times n}$$ (where T denotes transposed).

Then $S(t)$ satisfies the matrix Riccati equation

$$(6.42) \qquad \frac{dS}{dt} = FS + SF^T - SG^T(DD^T)^{-1}GS + CC^T, \qquad S(0) = E[X_0X_0^T]$$

and $\hat{X}_t$ satisfies

$$(6.43) \qquad d\hat{X}_t = F\hat{X}_t dt + SG^T(DD)^{-1}dN_t, \quad \hat{X}_0 = E[X_0]$$

or

$$(6.44) \qquad d\hat{X}_t = (F-SG^T(DD^T)^{-1}G)\hat{X}_t dt + SG^T(DD^T)^{-1}dZ_t \; ; \quad \hat{X}_0 = E[X_0].$$

The condition on $D(t) \in \mathbb{R}^{m\times r}$ is now that $D(t)D(t)^T$ is invertible for all t and that $(D(t)D(t)^T)^{-1}$ is bounded on every bounded interval.

A similar solution can be found for the more general situation

(6.45) (System) $\quad dX_t = [F_0(t) + F_1(t)X_t + F_2(t)Z_t]dt + C(t)dB_t$

(6.46) (Observations) $\quad dZ_t = [G_0(t) + G_1(t)X_t + G_2(t)Z_t]dt + D(t)dB_t$,

where $X_t \in \mathbb{R}^n$, $Y_t \in \mathbb{R}^m$ and B_t is $n+m$-dimensional Brownian motion, with appropriate dimensions on the matrix coefficients. (See Kallianpur [1], who also treats the non-linear case).

For the solution of linear filtering problems governed by more general processes than Brownian motion (processes with orthogonal increments) see Davis [1].

For application of filtering theory to space navigation etc. see Bucy and Joseph [1], Jazwinski [1] and the references in these books.

VII. Diffusions: Basic Properties

Suppose we want to describe the motion of a small particle suspended in a moving liquid, subject to random molecular bombardments. If $b(t,x) \in \mathbb{R}^3$ is the velocity of the fluid at the point x at time t, then a reasonable mathematical model for the position X_t of the particle at time t would be a stochastic differential equation of the form

(7.1) $$\frac{dX_t}{dt} = b(t,X_t) + \sigma(t,X_t)W_t,$$

where $W_t \in \mathbb{R}^3$ denotes "white noise" and $\sigma(t,x) \in \mathbb{R}^{3\times 3}$. The Ito interpretation of this equation is

(7.2) $$dX_t = b(t,X_t)dt + \sigma(t,X_t)dB,$$

where B_t is 3-dimensional Brownian motion, and similarly (with a correction term added to b) for the Stratonovich interpretation (see (6.2)).

In a stochastic differential equation of the form

(7.3) $$dX_t = b(t,X_t)dt + \sigma(t,X_t)dB_t,$$

where $X_t \in \mathbb{R}^n$, $b(t,x) \in \mathbb{R}^n$, $\sigma(t,x) \in \mathbb{R}^{n\times m}$ and B_t is m-dimensional Brownian motion, we will call b the <u>drift coefficient</u> and $\frac{1}{2}\sigma\sigma^T$ the <u>diffusion coefficient</u> (see Theorem 7.8).

Thus the solution of a stochastic differential equation may be thought of as the mathematical description of the motion of a small particle in a moving fluid: Therefore such stochastic processes are called <u>(Ito) diffusions</u>.

In this chapter we establish some of the most basic properties and results about Ito diffusions:

(A) The Markov property.
(B) The strong Markov property.
(C) The generator A of X_t expressed in terms of b and σ.
(D) The Dynkin formula.
(E) The characteristic operator.

This will give us the necessary background for the applications in Chapters IX, X and XI.

(A) The Markov property

DEFINITION 7.1. A (time-homogeneous) Ito diffusion is a stochastic process $X_s(\omega) = X(s,\omega) : [0,\infty) \times \Omega \to \mathbb{R}^n$ satisfying a stochastic differential equation of the form

(7.4) $$dX_s = b(X_s)ds + \sigma(X_s)dB_s, \quad s > t$$
$$X_t = x$$

where B_s is m-dimensional Brownian motion and $b: \mathbb{R}^n \to \mathbb{R}^n$, $\sigma: \mathbb{R}^n \to \mathbb{R}^{n\times m}$ satisfy the conditions in Theorem 5.5:

(7.5) $$|b(x)| + |\sigma(x)| \leq C(1+|x|), \quad \text{where} \quad |\sigma|^2 = \Sigma|\sigma_{ij}|^2$$

(7.6) $$|b(x)-b(y)| + |\sigma(x)-\sigma(y)| \leq D|x-y| \; ; \; x, y \in \mathbb{R}^n.$$

We will denote the (unique) solution of (7.4) by $X_s = X_s^{t,x}$; $s \geq t$. If $t=0$ we write X_s^x for $X_s^{0,x}$. Note that we have assumed in (7.4) that b and σ do not depend on t but on x only. We shall see later (Chapter X) that the general case can be reduced to this situation. The resulting process $X_t(\omega)$ will have the property of being time-homogeneous, in the following sense:

Note that

$$X_{t+h}^{t,x} = x + \int_t^{t+h} b(X_u^{t,x})du + \int_t^{t+h} \sigma(X_u^{t,x})dB_u$$

$$= x + \int_0^h b(X_{t+v}^{t,x})dv + \int_0^h \sigma(X_{t+v}^{t,x})d\tilde{B}_v, \quad (u=t+v)$$

where $\tilde{B}_v = B_{t+v} - B_t$; $v \geq 0$. On the other hand of course

$$X_h^{0,x} = x + \int_0^h b(X_v^{0,x})dv + \int_0^h \sigma(X_v^{0,x})dB_v.$$

Since $\{\tilde{B}_v\}_{v\geq 0}$ and $\{B_v\}_{v\geq 0}$ have the same P^0-distributions, it follows by weak uniqueness (see (5.27)) of the solution of the stochastic differential equation

$$dX_t = b(X_t)dt + \sigma(X_t)dB_t; \; X_0 = x$$

that

$$\{X_{t+h}^{t,x}\}_{h\geq 0} \quad \text{and} \quad \{X_h^{0,x}\}_{h\geq 0}$$

have the same P^0-distributions, i.e. $\{X_t\}_{t\geq 0}$ is time-homogeneous.

We now introduce the probability laws Q^x of $\{X_t\}_{t\geq 0}$, for $x \in \mathbb{R}^n$. Intuitively, Q^x gives the distribution of $\{X_t\}_{t\geq 0}$ assuming that $X_0 = x$. To express this mathematically, we let $\mathcal{M}$ be the σ-algebra (of subsets of Ω) generated by the random variables $\omega \to X_t(\omega) = X_t^x(\omega)$, where $t \geq 0$, $x \in \mathbb{R}^n$.

Define Q^x on the members of $\mathcal{M}$ by

(7.7) $\quad Q^x[X_{t_1} \in E_1, \ldots, X_{t_k} \in E_k] = P^0[X_{t_1}^x \in E_1, \ldots, X_{t_k}^x \in E_k]$ where

$E_i \subset \mathbb{R}^n$ are Borel sets; $1 \leq i \leq k$.

As before we let $\mathcal{F}_t$ be the σ-algebra generated by $\{B_s;\ s\leq t\}$. Similarly we let $\mathcal{M}_t$ be the σ-algebra generated by $\{X_s;\ s\leq t\}$. We have established earlier (see Theorem 5.5) that X_t is measurable with respect to $\mathcal{F}_t$, so $\mathcal{M}_t \subseteq {}_t$.

We now prove that X_t satisfies the important <u>Markov property</u>: The future behaviour of the process given what has happened up to time t is the same as the behaviour obtained when starting the process at X_t. The precise mathematical formulation of this is the following:

<u>THEOREM 7.2.</u> <u>(The Markov property for Ito diffusions)</u>
Let f be a bounded Borel function from $\mathbb{R}^n$ to $\mathbb{R}$. Then, for $t, h \geq 0$

(7.8) $$E^x[f(X_{t+h})|\mathcal{F}_t]_{(\omega)} = E^{X_t(\omega)}[f(X_h)].$$

(See Appendix B for definition and basic properties of conditional expectation). Here and in the following E^x denotes the expectation wrt. the probability measure Q^x. The right hand side means the function $E^y[f(X_h)]$ evaluated at $y = X_t(\omega)$.

<u>Proof.</u> Since

$$X_s(\omega) = X_t(\omega) + \int_t^s b(X_u)du + \int_t^s \sigma(X_u)dB_u,$$

we have by uniqueness

$$X_s(\omega) = X_s^{t,X_t}(\omega).$$

In other words, if we define

$$F(x,t,s,\omega) = X_s^{t,x}(\omega) \quad \text{for} \quad s \geq t,$$

we have

(7.9) $$X_s(\omega) = F(X_t,t,s,\omega)\ ; \ s \geq t.$$

Note that $\omega \to F(x,t,s,\omega)$ is independent of $\mathcal{F}_t$. Using (7.9) we may rewrite (7.8) as

(7.10) $$E[f(F(X_t,t,t+h,\omega))|\mathcal{F}_t] = E[f(F(x,0,h,\omega))]_{x=X_t}.$$

where E denotes the expectation wrt. the measure P^0.

Put $g(x,\omega) = f \circ F(x,t,t+h,\omega)$ and approximate g pointwise boundedly by functions on the form

$$\sum_{k=1}^{m} \phi_k(x)\psi_k(\omega).$$

(It follows from (5.22) that the map $x \to g(x,\omega)$ is continuous into $L^2(\Omega)$ and therefore $(x,\omega) \to g(x,\omega)$ is measurable).

Using the properties of conditional expectation (see Appendix B) we get

$$E[g(X_t,\omega)|\mathcal{F}_t] = E[\lim \Sigma\ \phi_k(X_t)\psi_k(\omega)|\mathcal{F}_t]$$

$$= \lim \Sigma\ \phi_k(X_t) \cdot E[\psi_k(\omega)|\mathcal{F}_t]$$

$$= \lim \Sigma\ E[\phi_k(y)\psi_k(\omega)|\mathcal{F}_t]_{y=X_t}$$

$$= E[g(y,\omega)|\mathcal{F}_t]_{y=X_t} = E[g(y,\omega)]_{y=X_t}$$

Therefore, since $\{X_t\}$ is time-homogeneous,

$$E[f(F(X_t,t,t+h,\omega))|\mathcal{F}_t] = E[f(F(y,t,t+h,\omega))]_{y=X_t}$$

$$= E[f(F(y,0,h,\omega))]_{y=X_t}$$

which is (7.10).

REMARK. Theorem 7.2 states that X_t is a Markov process wrt. the family of σ-algebras $\{\mathcal{F}_t\}_{t\geq 0}$. Note that since $\mathcal{M}_t \subseteq \mathcal{F}_t$ this implies that X_t is also a Markov process wrt. the σ-algebras $\{\mathcal{M}_t\}_{t\geq 0}$. This follows from Theorem B.3 and Theorem B.2 c) Appendix B):

$$E^x[f(X_{t+h})|\mathcal{M}_t] = E^x[E^x[f(X_{t+h})|\mathcal{F}_t]|\mathcal{M}_t]$$

$$= E^x[E^{X_t}[f(X_h)]|\mathcal{M}_t]$$

$$= E^{X_t}[f(X_h)]$$

since $E^{X_t}[f(X_h)]$ is $\mathcal{M}_t$-measurable.

(B) The strong Markov property

Roughly, the strong Markov property states that a relation of the form (7.8) continues to hold if the time t is a replaced by a random time $\tau(\omega)$ of a more general type called stopping time:

DEFINITION 7.3. Let $\{\mathcal{N}_t\}$ be an increasing family of σ-algebras (of subsets of Ω). A function $\tau : \Omega \to [0,\infty]$ is called a (strict) stopping time wrt. $\mathcal{N}_t$ if

$$\{\omega;\ \tau(\omega) \leq t\} \in \mathcal{N}_t, \quad \text{for all } t \geq 0.$$

In other words, it should be possible to decide whether or not $\tau \leq t$ has occurred on the basis of the knowledge of $\mathcal{N}_t$.

EXAMPLE 7.4. Let $U \subset \mathbb{R}^n$ be open. Then the first exit time

$$\tau_U = \inf\{t>0;\ X_t \notin U\}$$

is a stopping time wrt. $\mathcal{M}_t$, since

$$\{\omega;\ \tau_U \leq t\} = \bigcap_m \bigcup_{\substack{r\in Q\\ r<t}} \{\omega;\ X_r \notin K_m\} \in \mathcal{M}_t$$

where $\{K_m\}$ is an increasing sequence of closed sets such that $U = \bigcup_m K_m$.

More generally, if $H \subset \mathbb{R}^n$ is any set we define the first exit time from H, τ_H, as follows

$$\tau_H = \inf\{t>0;\ X_t \notin H\}.$$

If we include the sets of measure 0 in $\mathcal{M}_t$ (which we do) then the family $\{\mathcal{M}_t\}$ is right-continuous i.e. $\mathcal{M}_t = \mathcal{M}_{t+}$, where $\mathcal{M}_{t+} = \bigcap_{s>t} \mathcal{M}_s$ (see Chung [1], Theorem 2.3.4., p.61) and therefore τ_H is a stopping time for any Borel set H (see Dynkin [3], 4.5.C.e), p.111).

DEFINITION 7.5. Let τ be a stopping time wrt. $\{\mathcal{N}_t\}$ and let $\mathcal{N}_\infty$ be the smallest σ-algebra containing $\mathcal{N}_t$ for all $t \geq 0$. Then the σ-algebra $\mathcal{N}_\tau$ consists of all sets $N \in \mathcal{N}_\infty$ such that

$$N \cap \{\tau \leq t\} \in \mathcal{N}_t \quad \text{for all } t \geq 0.$$

In the case when $\mathcal{N}_t = \mathcal{M}_t$, an alternative and more intuitive description is:

(7.11) $\quad \mathcal{M}_\tau = \sigma\text{-algebra generated by } \{X_{\min(s,\tau)};\ s \geq 0\}.$

(See Rao [1] p. 2.15 or Stroock and Varadhan [1], Lemma 1.3.3, p.33). Similarly, if $\mathcal{N}_t = \mathcal{F}_t$, we get

$$\mathcal{F}_\tau = \sigma\text{-algebra generated by } \{B_{\min(s,\tau)};\ s \geq 0\}.$$

THEOREM 7.6. (The strong Markov property for Ito diffusions)
Let f be a bounded Borel function on $\mathbb{R}^n$, τ a stopping time wrt. $\mathcal{F}_t$, $\tau < \infty$ a.s. Then

$$E^x[f(X_{\tau+h})|\mathcal{F}_\tau] = E^{X_\tau}[f(X_h)] \quad \text{for all } h \geq 0. \tag{7.12}$$

Proof. We try to imitate the proof of the Markov property (Theorem 7.2). For a.a. ω we have

$$X_{\tau+h}(\omega) = X_\tau(\omega) + \int_{\tau(\omega)}^{\tau(\omega)+h} b(X_u)du + \int_{\tau(\omega)}^{\tau(\omega)+h} \sigma(X_u)dB_u. \tag{7.13}$$

By the strong Markov property for Brownian motion (Gihman and Skorohod [1], p. 30) the process

$$\tilde{B}_v = B_{\tau+v} - B_\tau\ ;\ v \geq 0$$

is again a Brownian motion and independent of $\mathcal{F}_\tau$.
Therefore, from (7.13)

$$X_{\tau+h}(\omega) = X_\tau(\omega) + \int_0^h b(X_{\tau+v})dv + \int_0^h \sigma(X_{\tau+v})d\tilde{B}_v.$$

So if we let $\tilde{F}(x,t,s,\omega)$ denote the unique solution X of

$$X_s = x + \int_t^s b(X_u)du + \int_t^s \sigma(X_u)d\tilde{B}_u.$$

we have $\omega \to \tilde{F}(x,t,s,\omega)$ independent of $\mathcal{F}_\tau$ and

$$X_{\tau+h} = \tilde{F}(X_\tau,0,h,\omega).$$

Just as in the proof of Theorem 7.2 this gives

$$E[f(\tilde{F}(X_\tau,0,h,\omega))|\mathcal{F}_\tau] = E[f(\tilde{F}(y,0,h,\omega))]_{y=X_\tau},$$

which is (7.12).

We now extend (7.12) to the following:
If $f_1,\dots,f_k$ are bounded Borel functions on $\mathbb{R}^n$, τ an $\mathcal{F}_t$-stopping time, $\tau < \infty$ a.s. then

$$E^x[f_1(X_{\tau+h_1})f_2(X_{\tau+h_2})\cdots f_k(X_{\tau+h_k})|\mathcal{F}_\tau] = E^{X_\tau}[f_1(X_{h_1})\cdots f_k(X_{h_k})] \tag{7.14}$$

for all $0 \leq h_1 \leq h_2 \leq \cdots \leq h_k$. This follows by induction: To illustrate the argument we prove it in the case $k=2$:

$$E^x[f_1(X_{\tau+h_1})f_2(X_{\tau+h_2})|\mathcal{F}_\tau]=E^x[E^x[f_1(X_{\tau+h_1})f_2(X_{\tau+h_2})|\mathcal{F}_{\tau+h_1}]|\ _\tau]$$

$$= E^x[f_1(X_{\tau+h_1})E^x[f_2(X_{\tau+h_2})|\mathcal{F}_{\tau+h_1}]|\mathcal{F}_\tau]$$

$$= E^x[f_1(X_{\tau+h_1})E^{X_{\tau+h_1}}[f_2(X_{h_2-h_1})]|\mathcal{F}_\tau]$$

$$= E^{X_\tau}[f_1(X_{h_1})E^{X_{h_1}}[f_2(X_{h_2-h_1})]]$$

$$= E^{X_\tau}[f_1(X_{h_1})E^x[f_2(X_{h_2})|\mathcal{F}_{h_1}]] = E^{X_\tau}[f_1(X_{h_1})f_2(X_{h_2})],$$ as claimed

Next we proceed to formulate the general version we need: Let $\mathcal{X}$ be the set of all real $\mathcal{M}$-measurable functions. For $t \geq 0$ we define the <u>shift operator</u>

$$\theta_t: \mathcal{X} \to \mathcal{X}$$

as follows:

If $\eta = g_1(X_{t_1})\cdots g_k(X_{t_k})$ (g_i Borel measurable, $t_i \geq 0$) we put

$$\theta_t\eta = g_1(X_{t_1+t})\cdots g_k(X_{t_k+t}).$$

Now extend in the natural way to all functions in $\mathcal{X}$ by taking limits of sums of such functions. Then it follows from (7.13) that

$$(7.15) \qquad E^x[\theta_\tau\eta|\mathcal{F}_\tau] = E^{X_\tau}[\eta]$$

for all stopping times τ and all bounded $\eta \in \mathcal{X}$, where

$$(\theta_\tau\eta)(\omega) = (\theta_t\eta)(\omega) \quad \text{if} \quad \tau(\omega) = t.$$

We will apply this to the following situation: Let $H \subset \mathbb{R}^n$ be measurable and let τ_H be the first exit time from H for an Ito diffusion X_t. Let α be another stopping time, g a bounded continuous function on $\mathbb{R}^n$ and put

$$\eta = g(X_{\tau_H})\,\chi_{\{\tau_H<\infty\}}, \quad \tau_H^\alpha = \inf\{t>\alpha;\ X_t \notin H\}$$

Then we have

$$(7.16) \qquad \theta_\alpha\eta \cdot \chi_{\{\alpha<\infty\}} = g(X_{\tau_H^\alpha})\chi_{\{\tau_H^\alpha<\infty\}}$$

To prove (7.16) we approximate η by functions $\eta^{(k)}$; $k = 1,2,\ldots,$ of the form

$$\eta^{(k)} = \sum_j g(X_{t_j})\,\chi_{[t_j,t_{j+1})}(\tau_H)\ , \quad t_j=j\cdot 2^{-k},\ j=0,1,2,\ldots$$

Now $\theta_t \chi_{[t_j, t_{j+1})}(\tau_H) = \theta_t \chi_{\{\forall r \in (0,t_j) X_r \in H \ \& \ \exists s \in [t_j, t_{j+1}) X_s \notin H\}}$

$= \chi_{\{\forall r \in (0,t_j) X_{r+t} \in H \ \& \ \exists s \in [t_j, t_{j+1}) X_{s+t} \notin H\}}$

$= \chi_{\{\forall u \in (t, t_{j+1}) X_u \in H \ \& \ \exists v \in [t_j+t,\ t_{j+1}+t) X_v \notin H\}} = \chi_{[t_j+t,\ t_{j+1}+t)}(\tau_H^t)$.

So we see that

$$\theta_t \eta = \lim_k \theta_t \eta^{(k)} = \lim_k \sum_j g(X_{t_j+t})\, \chi_{[t_j+t,\ t_{j+1}+t)}(\tau_H^t)$$

$$= g(X_{\tau_H} t) \cdot \chi_{\{\tau_H^t < \infty\}}, \quad \text{which is (7.16).}$$

In particular, if $\alpha = \tau_G$ with $G \subset\subset H$ measurable, $\tau_H < \infty$ a.s. Q^x, then we have $\tau_H^\alpha = \tau_H$ and so

$$(7.17) \qquad \theta_{\tau_G} g(X_{\tau_H}) = g(X_{\tau_H}).$$

So if f is any bounded measurable function we obtain from (7.15) and (7.17) (define $\mu_x(F) = Q^x(X_{\tau_H} \in F)$ and approximate f in $L^1(\mu_x)$ by continuous functions g satisfying (7.17))

$$(7.18) \qquad E^x[f(X_{\tau_H})] = E^x[E^{X_{\tau_G}}[f(X_{\tau_H})]] = \int_{\partial G} E^y[f(X_{\tau_H})] \cdot Q^x[X_{\tau_G} \in dy]$$

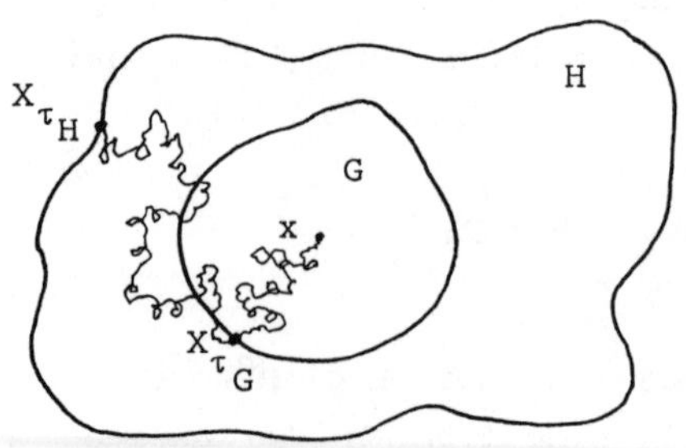

In other words, the expected value of f at X_{τ_H} when starting at x can be obtained by integrating the expected value when starting at $y \in \partial G$ with respect to the <u>hitting distribution</u> ("harmonic measure") of X on ∂G. This can be restated as follows:

Define the harmonic measure of X on ∂G, $\mu_x^{X,G}$, by

$$\mu_x^{X,G}(F) = Q^x[X_{\tau_G} \in F] \quad \text{for } F \subset \partial G$$

Then the function

$$\phi(x) = E^x[f(X_{\tau_H})]$$

satisfies <u>the mean value property</u>:

$$(7.19) \qquad \phi(x) = \int_{\partial G} \phi(y)\, d\mu_x^{X,G}(y),$$

for all Borel sets $G \subset\subset H$.

This is an important ingredient in our solution of the generalized Dirichlet problem in Chapter IX.

(C) The generator of an Ito diffusion

It is fundamental for many applications that we can associate a second order partial differential operator L to an Ito diffusion X_t. The basic connection between L and X_t is that L is the generator of the process X_t.

DEFINITION 7.7. Let $\{X_t\}$ be a time-homogeneous Ito diffusion in $\mathbb{R}^n$. The (infinitesimal) generator A of X_t is defined by

$$(Af)(x) = \lim_{t\downarrow 0} \frac{E^x[f(X_t)] - f(x)}{t} \quad ; \; x \in \mathbb{R}^n.$$

The set of functions $f : \mathbb{R}^n \to \mathbb{R}$ such that the limit exists at x is denoted by $\mathcal{D}_A(x)$, while $\mathcal{D}_A$ denotes the set of functions for which the limit exists for all $x \in \mathbb{R}^n$.

To find the relation between A and coefficients b, σ in the stochastic differential equation (7.4) defining X_t we need the following result, which is useful in many connections:

LEMMA 7.8. Let $Y_t = Y_t^x$ be a stochastic integral of the form

$$Y_t^x(\omega) = x + \int_0^t u(s,\omega)ds + \int_0^t v(s,\omega)dB_s(\omega)$$

where u, v satisfy conditions (4.4), (4.5). Let $f \in C_0^2(\mathbb{R}^n)$ and let τ be a stopping time wrt. $\{\mathcal{F}_t\}$, $E[\tau] = \int \tau dP^0 < \infty$. Assume that $u(t,\omega)$ and $v(t,\omega)$ are bounded on the set of (t,ω) such that $Y(t,\omega)$ belongs to the support of f. Then

$$E^x[f(Y_\tau)] = f(x) + E^x\Big[\int_0^\tau \Big(\sum_i u_i(s,\omega)\frac{\partial f}{\partial x_i}(Y_s) +$$

$$\tfrac{1}{2}\sum_{i,j}(vv^T)_{i,j}\frac{\partial^2 f}{\partial x_i \partial x_j}(Y_s)\Big)ds\Big],$$

where E^x is the expectation wrt. the natural probability law R^x for Y_t starting at x:

$$R^x[Y_{t_1} \in E_1, \dots, Y_{t_k} \in E_k] = P^0[Y_{t_1}^x \in E_1, \dots, Y_{t_k}^x \in E_k], \quad E_i \text{ Borel sets.}$$

Proof. Put $Z = f(Y)$ and apply Ito's formula (To simplify the notation we suppress the index t and let $Y_1,\ldots,Y_n$ and $B_1,\ldots,B_m$ denote the coordinates of Y and B, respectively)

$$dZ = \sum_i \frac{\partial f}{\partial x_i}(Y)\, dY_i + \tfrac{1}{2} \sum_{i,j} \frac{\partial^2 f}{\partial x_i \partial x_j} (Y)\, (dY_i)\, (dY_j)$$

$$= \sum_i u_i \frac{\partial f}{\partial x_i}\, dt + \tfrac{1}{2} \sum_{i,j} \frac{\partial^2 f}{\partial x_i \partial x_j} (vdB)_i\, (vdB)_j$$

$$+ \sum_i \frac{\partial f}{\partial x_i} (vdB)_i .$$

Since
$$(vdB)_i \cdot (vdB)_j = (\sum_k v_{ik} dB_k)(\sum_n v_{jn} dB_n)$$
$$= (\sum_k v_{ik} v_{jk}) dt = (vv^T)_{ij} dt,$$

this gives

$$f(Y_t) = f(Y_0) + \int_0^t (\sum_i u_i \frac{\partial f}{\partial x_i} + \tfrac{1}{2} \sum_{i,j} (vv^T)_{ij} \frac{\partial^2 f}{\partial x_i \partial x_j})ds \tag{7.20}$$
$$+ \sum_{i,k} \int_0^t v_{ik} \frac{\partial f}{\partial x_i}\, dB_k .$$

Hence

$$E^x[f(Y_\tau)] = f(x) + E^x[\int_0^\tau (\sum_i u_i \frac{\partial f}{\partial x_i}(Y) + \tfrac{1}{2} \sum_{i,j} (vv^T)_{i,j} \frac{\partial^2 f}{\partial x_i \partial x_j}(Y))ds]$$

$$+ \sum_{i,k} (E^x \int_0^\tau v_{ik} \frac{\partial f}{\partial x_i}(Y) dB_k]) . \tag{7.21}$$

If g is a bounded Borel function, $g \le M$ say, then for all integers k we have

$$E^x[\int_0^{\tau\wedge k} g(Y_s) dB_s] = E^x[\int_0^k \chi_{\{s<\tau\}} g(Y_s) dB_s] = 0,$$

since $g(Y_s)$ and $\chi_{\{s<\tau\}}$ are both $\mathcal{F}_s$-measurable. Moreover

$$E^x[(\int_0^{\tau\wedge k} g(Y_s) dB_s)^2] = E^x[\int_0^{\tau\wedge k} g^2(Y_s) ds] \le M^2 E^x[\tau] < \infty .$$

Therefore the family $\{\int_0^{\tau\wedge k} g(Y_s) dB_s\}_k$ is <u>uniformly integrable</u> w.r.t. the measure R^x (see Appendix C). It follows that

$$0 = \lim_{k\to\infty} E^x[\int_0^{\tau\wedge k} g(Y_s) dB_s] = E^x[\lim_{k\to\infty} \int_0^{\tau\wedge k} g(Y_s) dB_s] = E^x[\int_0^\tau g(Y_s) dB_s].$$

Combining this with (7.21) we get Lemma 7.8.

This gives immediately the formula for the generator A for an Ito diffusion:

THEOREM 7.9. Let X_t be the Ito diffusion

$$dX_t = b(X_t)dt + \sigma(X_t)dB_t$$

If $f \in C^2(\mathbb{R}^n)$ is bounded with bounded first and second derivatives then $f \in \mathcal{D}_A$ and

$$(7.22) \qquad (Af)(x) = \sum_i b_i(x)\frac{\partial f}{\partial x_i} + \tfrac{1}{2}\sum_{i,j}(\sigma\sigma^T)_{i,j}(x)\frac{\partial^2 f}{\partial x_i \partial x_j}$$

Proof. This follows from Lemma 7.8 (with $\tau=t$) and the definition of A.

From now on we will, unless otherwise stated, let A denote the generator of the Ito diffusion X_t. We let L denote the differential operator given by the right hand side of (7.22). From Theorem 7.9 we know that A and L coincide on $C_0^2(\mathbb{R}^n)$.

(D) The Dynkin formula.

If we combine (7.21) and (7.22) we get:

THEOREM 7.10 (Dynkin's formula)

Let $f \in C_0^2(\mathbb{R}^n)$. Suppose τ is a stopping time, $E[\tau]<\infty$. Then

$$(7.23) \qquad E^x[f(X_\tau)] = f(x) + E^x\left[\int_0^\tau (Af)(X_s)ds\right]$$

REMARKS.

(i) Note that if τ is the first exit time of a bounded set, $E[\tau]<\infty$, then (7.23) holds for any function $f \in C^2$.

(ii) It is possible to prove the theorem under the weaker assumption that f is a bounded function in $\mathcal{D}_A$. See Dynkin [2], p. 133.

EXAMPLE 7.11. Consider n-dimensional Brownian motion $B=(B_1,\dots,B_n)$ starting at $a=(a_1,\dots,a_n) \in \mathbb{R}^n (n \geq 2)$ and assume $|a| < R$. What is the expected value of the first exit time τ_k of B from the ball

$$K = K_R = \{x \in \mathbb{R}^n;\ |x|<R\}\ ?$$

Choose an integer k and apply Dynkin's formula with $X=B$, $\tau=\sigma_k=\min(k,\tau_K)$, and $f \in C_0^2$ such that $f(x) = |x|^2$ for $|x| \leq R$:

$$E^a[f(B_{\sigma_k})] = f(a) + E^a[\int_0^{\sigma_k} \tfrac{1}{2}\Delta f(B_s)ds]$$

$$= |a|^2 + E^a[\int_0^{\sigma_k} n\cdot ds] = |a|^2 + n \cdot E^a[\sigma_k]$$

Hence $E^a[\sigma_k] \leq \frac{1}{n}(R^2-|a|^2)$ for all k. So letting $k \to \infty$ we conclude that $\tau_K = \lim \sigma_k < \infty$ a.s. and

(7.24) $$E^a[\tau_K] = \frac{1}{n}(R^2-|a|^2)$$

Next we assume that $|b| > R$. What is the probability that B starting at b over hits K?

Let α_k be the first exit time from the annulus

$$A_k = \{x;\ R<|x|<2^kR\};\quad k = 1,2,..$$

Let $f = f_{n,k}$ be a C^2 function with compact support such that, if $R \leq |x| \leq 2^kR$,

$$f(x) = \begin{cases} -\log|x| & \text{when } n = 2 \\ |x|^{2-n} & \text{when } n > 2 \end{cases}$$

Then, since $\Delta f = 0$ in A_k, we have by Dynkin's formula

(7.25) $$E^b[f(B_{\alpha_k})] = f(b) \quad \text{for all } k.$$

Put

$$p_k = P^b[|B_{\alpha_k}| = R],\ q_k = P^b[|B_{\alpha_k}| = 2^kR].$$

Let us now consider the two cases $n = 2$ and $n > 2$ separately:

<u>$n = 2$</u>. Then we get from (7.25)

(7.26) $$-\log R \cdot p_k - (\log R + k\cdot\log 2)q_k = -\log|b| \quad \text{for all } k$$

This implies that $q_k \to 0$ as $k \to \infty$, so that

(7.27) $$P^b[\tau_K < \infty] = 1,$$

i.e. Brownian motion is <u>recurrent</u> in $\mathbb{R}^2$. (See Port and Stone [1]).

<u>$n > 2$</u>. In this case (7.25) gives

$$p_k \cdot R^{2-n} + q_k \cdot (2^kR)^{2-n} = |b|^{2-n}$$

Since $0 \leq q_k \leq 1$ we get by letting $k \to \infty$

$$\lim_{k\to\infty} p_k = P^b[\tau_K < \infty] = \left(\frac{|b|}{R}\right)^{2-n},$$

i.e. Brownian motion is <u>transient</u> in $\mathbb{R}^n$ for $n > 2$.

(E) The characteristic operator

We now introduce an operator which is closely related to the generator A, but is more suitable in many situations, for example in the solution of the Dirichlet problem.

DEFINITION 7.12. Let $\{X_t\}$ be an Ito diffusion. The characteristic operator $\mathscr{A}$ of $\{X_t\}$ is defined by

$$(7.28) \qquad (\mathscr{A}f)(x) = \lim_{U \downarrow x} \frac{E^x[f(X_{\tau_U})]-f(x)}{E^x[\tau_U]},$$

where the U's are open sets decreasing to the point x and $\tau_U = \inf\{t>0;\ X_t \notin U\}$ is the first exit time from U for X_t. The set of functions f such that the limit (7.28) exists for all $x \in \mathbb{R}^n$ is denoted by $\mathscr{D}_{\mathscr{A}}$. If $E^x[\tau_U] = \infty$ for all open $U \ni x$, we define $(\mathscr{A}f)(x) = 0$.

It turns out that $\mathscr{D}_A \subseteq \mathscr{D}_{\mathscr{A}}$ always and that

$$Af = \mathscr{A}f \quad \text{for all} \quad f \in \mathscr{D}_A.$$

(See Dynkin [2], p. 143).

We will only need that $\mathscr{A}$ and A coincide on C^2. To obtain this we first clarify a property of exit times.

DEFINITION 7.13. A point $x \in \mathbb{R}^n$ is called a trap for $\{X_t\}$ if

$$Q^x(\{X_t = x \quad \text{for all} \quad t\}) = 1.$$

In other words, x is trap if and only if $\tau_x = \infty$ a.s. Q^x, where

$$\tau_x = \inf\{t>0;\ X_t \neq x\}$$

is the first exit time from the set $\{x\}$. For example, if $b(x_0)=\sigma(x_0)=0$, then x_0 is a trap for X_t (by strong uniqueness of X_t).

LEMMA 7.14. If x is not a trap for X_t, then there exists an open set $U \ni x$ such that

$$E^x[\tau_U] < \infty.$$

Proof. See Lemma 5.5 p. 139 in Dynkin [2].

THEOREM 7.15. Let $f \in C^2$. Then $f \in \mathcal{D}_{\mathscr{A}}$ and

$$(7.29) \quad \mathscr{A}f = Lf \overset{\text{def}}{=} \sum_i b_i \frac{\partial f}{\partial x_i} + \frac{1}{2} \sum_{i,j} (\sigma \sigma^T)_{ij} \frac{\partial^2 f}{\partial x_i \partial x_j}$$

Proof. If x is a trap for $\{X_t\}$ then $(\mathscr{A}f)(x) = (Af)(x) = 0$. If x is not a trap, choose a bounded open set $U \ni x$ such that $E^x[\tau_U] < \infty$. Then by Dynkin's formula 7.10 (and the following Remark (i)), writing $\tau_U = \tau$

$$\left| \frac{E^x[f(X_\tau)] - f(x)}{E^x[\tau]} - (Lf)(x) \right| = \frac{\left| E^x[\int_0^\tau \{(Lf)(X_s) - (Lf)(x)\} ds] \right|}{E^x[\tau]}$$

$$\leq \sup_{y \in U} |(Lf)(x) - (Lf)(y)| \to 0 \quad \text{as } U \downarrow x,$$

since Lf is a continuous function.

REMARK. We have now obtained that an Ito diffusion is a continuous, strong Markov process such that the domain of definition of its characteristic operator includes C^2. Thus an Ito diffusion is a diffusion in the sense of Dynkin [2].

EXAMPLE 7.16. The n-dimensional Brownian motion is of course the solution of the stochastic differential equation

$$dX_t = dB_t,$$

i.e. we have $b = 0$ and $\sigma = I_n$, the n-dimensional identity matrix. So the generator of B_t is

$$Af = \tfrac{1}{2} \sum \frac{\partial^2 f}{\partial x_i^2},$$

i.e. $A = \frac{1}{2}\Delta$.

EXAMPLE 7.17. (The graph of Brownian motion). Let B denote 1-dimensional Brownian motion and let $X = \begin{bmatrix} X_1 \\ X_2 \end{bmatrix}$ be the solution of the stochastic differential equation

$$\begin{cases} dX_1 = dt & X_1(0) = t_0 \\ dX_2 = dB & X_2(0) = x_0 \end{cases}$$

i.e.

$$dX = b\,dt + \sigma\,d\tilde{B}, \quad X(0) = \begin{bmatrix} t_0 \\ x_0 \end{bmatrix}, \quad \text{with}$$

$b = \begin{bmatrix} 1 \\ 0 \end{bmatrix}$, $\sigma = \begin{bmatrix} 0 & 0 \\ 0 & 1 \end{bmatrix}$, $\tilde{B}$ a 2-dimensional Brownian motion. In other words, X may be regarded as the graph of Brownian motion. The characteristic operator $\mathscr{A}$ of X coincide by Theorem 7.13 on C^2 with the differential operator L given by

$$Lf = \frac{\partial f}{\partial t} + \tfrac{1}{2} \frac{\partial^2 f}{\partial x^2} \; ; \; f = f(t,x) \in C^2.$$

EXAMPLE 7.18. (Brownian motion on the unit circle). The characteristic operator of the process $Y = \begin{bmatrix} Y_1 \\ Y_2 \end{bmatrix}$ from Example 5.4. satisfying the stochastic equations (5.17)

$$\begin{cases} dY_1 = -\tfrac{1}{2} Y_1 dt - Y_2 dB \\ dY_2 = -\tfrac{1}{2} Y_2 dt + Y_1 dB \end{cases}$$

is

$$f(y_1,y_2) = \tfrac{1}{2} \left[y_2^2 \frac{\partial^2 f}{\partial y_1^2} - 2y_1 y_2 \frac{\partial^2 f}{\partial y_1 \partial y_2} + y_1^2 \frac{\partial^2 f}{\partial y_2^2} - y_1 \frac{\partial f}{\partial y_1} - y_2 \frac{\partial f}{\partial y_2} \right].$$

This is because $dY = -\tfrac{1}{2}Y dt + KY\, dB$, where

$$K = \begin{bmatrix} 0 & -1 \\ 1 & 0 \end{bmatrix}$$

so that

$$dY = b(Y)dt + \sigma(Y)dB$$

with

$$b(y_1,y_2) = \begin{bmatrix} -\tfrac{1}{2}y_1 \\ -\tfrac{1}{2}y_2 \end{bmatrix}, \quad \sigma(y_1,y_2) = \begin{bmatrix} -y_2 \\ y_1 \end{bmatrix} \quad \text{and}$$

$$a = \tfrac{1}{2}\sigma\sigma^T = \tfrac{1}{2} \begin{bmatrix} y_2^2 & -y_1 y_2 \\ -y_1 y_2 & y_1^2 \end{bmatrix}.$$

EXAMPLE 7.19. Let D be an open subset of $\mathbb{R}^n$ such that $\tau_D < \infty$ a.s. Q^x for all x. Let ϕ be a bounded, measurable function on ∂D and define

$$\tilde{\phi}(x) = E^x[\phi(X_{\tau_D})]$$

($\tilde{\phi}$ is called the <u>X-harmonic extension of ϕ</u>). Then if U is open, $x \in U \subset D$, we have by (7.18) that

$$E^x[\tilde{\phi}(X_{\tau_U})] = E^x[E^{X_{\tau_U}}[\phi(X_{\tau_D})]] = E^x[\phi(X_{\tau_D})] = \tilde{\phi}(x)$$

So $\tilde{\phi} \in \mathscr{D}_{\mathscr{A}}$ and

$$\mathscr{A}\tilde{\phi} = 0 \quad \text{in } D,$$

in spite of the fact that in general $\tilde{\phi}$ need not even be continuous in D (See Example 9.7).

VIII. Other Topics in Diffusion Theory

In this chapter we study other important topics in diffusion theory. While not strictly necessary for the remaining chapters, these topics are central in the theory of stochastic analysis and essential for further applications. The following topics will be treated:

A) Kolmogorov's backward equation. The resolvent.

B) The Feynman-Kac formula. Killing.

C) The martingale problem.

C) When is a stochastic integral a diffusion?

E) Random time change.

F) The Cameron-Martin-Girsanov formula.

A) Kolmogorov's backward equation. The resolvent.

If we choose $\tau=t$ in Dynkin's formula (7.23) we see that

$$u(t,x) = E^x[f(X_t)]$$

is differentiable with respect to t and

(8.1) $$\frac{\partial u}{\partial t} = E^x[(Af)(X_t)], \quad \text{for all} \quad f \in C_0^2.$$

It turns out that on the right hand side of (8.1) can be expressed in terms of u also:

THEOREM 8.1 (Kolmogorov's backward equation)

Let $f \in C_0^2(\mathbb{R}^n)$. Define

$$u(t,x) = E^x[f(X_t)]$$

Then

(8.2) $$\frac{\partial u}{\partial t} = Au,$$

where the right hand side is to be interpreted as A applied to the function $x \to u(t,x)$.

Proof. Let $g(x) = u(t,x)$. Then since $t \to u(t,x)$ is differentiable we have

$$\frac{E^x[g(X_r)]-g(x)}{r} = \frac{1}{r} \cdot E^x[E^{X_r}[f(X_t)] - E^x[f(X_t)]]$$

$$= \frac{1}{r} \cdot E^x[E^x[f(X_{t+r})|\mathcal{F}_r] - E^x[f(X_t)|\mathcal{F}_r]]$$

$$= \frac{1}{r} \cdot E^x[f(X_{t+r}) - f(X_t)]$$

$$= \frac{u(t+r,x)-u(t,x)}{r} \to \frac{\partial u}{\partial t} \quad \text{as} \quad r \downarrow 0.$$

Hence

$$Au = \lim_{r \downarrow 0} \frac{E^x[g(X_r)]-g(x)}{r} \quad \text{exists and} \quad \frac{\partial u}{\partial t} = Au, \quad \text{as asserted.}$$

REMARK. If we introduce the operator $Q_t: f \to E^{\cdot}[f(X_t)]$ then we have $u(x,t) = (Q_t f)(x)$ and we may rewrite (8.1) and (8.2) as follows:

$$(8.1)' \qquad \frac{d}{dt}(Q_t f) = Q_t(Af)$$

$$(8.2)' \qquad \frac{d}{dt}(Q_t f) = A(Q_t f)$$

Thus the equivalence of (8.1) and (8.2) amounts to saying that the operators Q_t and A commute, in some sense. Arguing formally, it is tempting to say that the solution of (8.1)' and (8.2)' is

$$Q_t = e^{tA}$$

and therefore $Q_t A = AQ_t$. However, this argument would require a further explanation, because in general A is an unbounded operator.

It is an important fact that the generator A of an Ito diffusion always has an inverse, at least if a positive multiple of the identity is substracted from A. This inverse can be expressed explicitly in terms of the diffusion X_t:

DEFINITION 8.2. For $\alpha>0$ and $g \in C_b(\mathbb{R}^n)$ we define the resolvent operator R_α by

$$(8.3) \qquad R_\alpha g(x) = E^x[\int_0^\infty e^{-\alpha t} g(X_t)dt]$$

LEMMA 8.3. $R_\alpha g$ is a bounded continuous function.

Proof. Since $R_\alpha g(x) = \int_0^\infty e^{-\alpha t} E^x[g(X_t)]dt$, we see that Lemma 8.3 is a direct consequence of the next result:

LEMMA 8.4. An Ito diffusion X_t is Feller-continuous, i.e.

$$u(x)=E^x[g(X_t)]$$

is a continuous function of x, for all bounded continuous functions g and all $t\geq 0$. If g is non-negative and continuous then $u(x)$ is lower semicontinous.

Proof. By (5.22) we have

$$E[|X_t^x - X_t^y|^2] \leq |y-x|^2 C(t),$$

where $C(t)$ does not depend on x and y. Let $\{y_n\}$ be a sequence of points converging to x. Then

$$X_t^{y_n} \to X_t^x \quad \text{in} \quad L^2(\Omega, P^0) \quad \text{as} \quad n\to\infty.$$

So, by taking a subsequence $\{z_n\}$ of $\{y_n\}$ we obtain that

$$X_t^{z_n}(\omega) \to X_t^x(\omega) \qquad \text{for a.a.}\ \omega \in \Omega.$$

If g is bounded and continuous then by bounded convergence

$$u(z_n) = E[g(X_t^{z_n})] \to E[g(X_t^x)] = u(x)$$

Therefore every sequence $\{y_n\}$ converging to x has a subsequence $\{z_n\}$ such that $u(z_n) \to u(x)$. That proves that u is continuous. If g is non-negative and continuous put $g_n = g\wedge n$. Then by monotone convergence $u(x) = E^x[g(X_t)] = \lim\limits_{n\to\infty} E^x[g_n(X_t)]$, so u is a limit of an increasing sequence of continuous functions. Therefore u is lower semicontinuous.

We now prove that R_α and $\alpha - A$ are inverse operators:

THEOREM 8.5.

a) If $f\in C_0^2(\mathbb{R}^n)$ then

$$R_\alpha(\alpha - A)f = f \qquad \text{for all} \quad \alpha > 0.$$

b) If $g\in C_b(\mathbb{R}^n)$ then $R_\alpha g \in \mathcal{D}_A$ and

$$(\alpha - A)R_\alpha g = g \qquad \text{for all} \quad \alpha > 0.$$

Proof. a) If $f\in C_0^2(\mathbb{R}^n)$ then by Dynkin's formula

$$R_\alpha(\alpha-A)f(x) = (\alpha R_\alpha f - R_\alpha Af)(x) = \alpha\int_0^\infty e^{-\alpha t}E^x[f(X_t)]dt - \int_0^\infty e^{-\alpha t}E^x[Af(X_t)]dt$$

$$= \Big|_0^\infty -e^{-\alpha t}E^x[f(X_t)] + \int_0^\infty e^{-\alpha t}\frac{d}{dt}E^x[f(X_t)]dt - \int_0^\infty e^{-\alpha t}E^x[Af(X_t)]dt$$

$$= E^x[f(X_0)] = f(x)$$

b) If $g\in C_b(\mathbb{R}^n)$ then by the strong Markov property

$$E^x[R_\alpha g(X_t)] = E^x[E^{X_t}[\int_0^\infty e^{-\alpha s}g(X_s)ds]]$$

$$= E^x[E^x[\theta_t(\int_0^\infty e^{-\alpha s}g(X_s)ds)|\mathcal{F}_t]] = E^x[E^x[\int_0^\infty e^{-\alpha s}g(X_{t+s})ds|\mathcal{F}_t]]$$

$$= E^x[\int_0^\infty e^{-\alpha s}g(X_{t+s})ds] = \int_0^\infty e^{-\alpha s}\, E^x[g(X_{t+s})]ds$$

Therefore, since $u(s,x)=E^x[g(X_s)]$ is s-differentiable,

$$\frac{1}{t}\,(E^x[R_\alpha g(X_t)]-R_\alpha g(x)) = \int_0^\infty e^{-\alpha s}\;\frac{u(t+s,x)\,-\,u(s,x)}{t}\,ds$$

$$\to \int_0^\infty e^{-\alpha s}\,\frac{\partial}{\partial s}\,u(s,x)ds \quad \text{as} \quad t \to 0.$$

Hence $R_\alpha g \in \mathcal{D}_A$ and

$$(\alpha-A)R_\alpha g(x) = \int_0^\infty \alpha e^{-\alpha s}u(s,x)ds - \int_0^\infty e^{-\alpha s}\,\frac{\partial}{\partial s}\,u(s,x)ds$$

$$= \Big|_0^\infty - e^{-\alpha s}u(s,x) = u(0,x) = g(x).$$

B) The Feynman-Kac formula. Killing

With a little harder work we can obtain the following useful generalization of Kolmogorov's backward equation:

THEOREM 8.6 (The Feynman-Kac formula)

Let $f \in C_0^2(\mathbb{R}^n)$ and put

(8.4) $$v(t,x) = E^x[\exp(-\int_0^t q(X_s)ds)\; f(X_t)],$$

where $q(x)$ is bounded and continuous. Then

(8.5) $$\frac{\partial v}{\partial t} = Av - q\,v.$$

Proof. Let $Y_t=f(X_t)$, $Z_t=\exp(-\int_0^t q(X_s)ds)$. Then dY_t is given by (7.20) and

$$dZ_t = -Z_t q(X_t)dt$$

So

$$d(Y_tZ_t) = Y_t dZ_t + Z_t dY_t, \quad \text{since} \quad dZ_t \cdot dY_t = 0.$$

Note that since $Y_t Z_t$ is a stochastic integral it follows from Lemma 7.8 that $v(t,x) = E^x[Y_t Z_t]$ is differentiable wrt. t.

Therefore

$$\frac{1}{r}(E^x[v(t,X_r)] - v(t,x)) = \frac{1}{r} E^x[E^{X_r}[Z_t f(X_t)] - E^x[Z_t f(X_t)]]$$

$$= \frac{1}{r} E^x[E^x[f(X_{t+r})\exp(-\int_0^t q(X_{s+r})ds)|\mathcal{F}_r] - E^x[Z_t f(X_t)|\mathcal{F}_r]]$$

$$= \frac{1}{r} E^x[Z_{t+r}\cdot\exp(\int_0^r q(X_s)ds)f(X_{t+r}) - Z_t f(X_t)]$$

$$= \frac{1}{r} E^x[f(X_{t+r})Z_{t+r} - f(X_t)Z_t] + \frac{1}{r} E^x[f(X_{t+r})Z_{t+r}\cdot\exp(\int_0^t q(X_s)ds) - 1)]$$

$$\to \frac{\partial}{\partial t} v(t,x) + q(x)\, v(t,x) \quad \text{as} \quad r \to 0,$$

because

$$\frac{1}{r} f(X_{t+r})Z_{t+r}(\exp(\int_0^r q(X_s)ds - 1) \to f(X_t)Z_t q(X_0)$$

pointwise boundedly. That completes the proof.

REMARK (About killing a diffusion)
In Theorem 7.9 we have seen that the generator of an Ito diffusion X_t given by

(8.6) $\quad dX_t = b(X_t)dt + \sigma(X_t)dB_t$

is a partial differential operator L of the form

(8.7) $\quad Lf = \sum a_{ij} \frac{\partial^2 f}{\partial x_i \partial x_j} + \sum b_i \frac{\partial f}{\partial x_i}$

where $[a_{ij}] = \frac{1}{2}\sigma\sigma^T$, $b = [b_i]$. It is natural to ask if one can also find processes whose generator has the form

(8.8) $\quad Lf = \sum a_{ij} \frac{\partial^2 f}{\partial x_i \partial x_j} + \sum b_i \frac{\partial f}{\partial x_i} - cf,$

where $c(x)$ is a bounded and continuous function.

If $c(x) \geq 0$ the answer is yes and a process $\tilde{X}_t$ with generator (8.8) is obtained by killing X_t at a certain (killing) time ζ. By this we mean that there exists a random time ζ such that if we put

(8.9) $\quad \tilde{X}_t = X_t \quad \text{if} \quad t < \zeta$

and leave $\tilde{X}_t$ undefined if $t \geq \zeta$ (alternatively, put $\tilde{X}_t = \partial$ if $t \geq \zeta$, where $\partial \notin \mathbb{R}^n$ is some "coffin" state), then $\tilde{X}_t$ is also a strong Markov process and

$$(8.10) \qquad E^x[f(\tilde{X}_t)] = E^x[f(X_t),\ t<\zeta] = E^x[f(X_t)\cdot e^{-\int_0^t c(X_s)ds}]$$

for all bounded continuous functions f on $\mathbb{R}^n$.

Let $v(t,x)$ denote the right hand side of (8.10) with $f \in C_0^2(\mathbb{R}^n)$. Then

$$\lim_{t\to 0} \frac{E^x[f(\tilde{X}_t)]-f(x)}{t} = \frac{\partial}{\partial t} v(t,x)_{t=0} = (Av-cv)_{t=0} = Af(x)-c(x)f(x),$$

by the Feynman-Kac formula.

So the generator of $\tilde{X}_t$ is (8.8), as required. The function $c(x)$ can be interpreted as the <u>killing rate</u>:

$$c(x) = \lim_{t\downarrow 0} \frac{1}{t} Q^x[X_t \text{ is killed in the time interval } (0,t)]$$

Thus by applying such a killing procedure we can come from the special case $c = 0$ in (8.7) to the general case (8.8) with $c(x) \geq 0$. Therefore, for many purposes it is enough to consider the equation (8.7).

In the function $c(x) \geq 0$ is given, an explicit construction of the killing time ζ such that (8.10) holds can be found in Karlin and Taylor [1], p. 314. For a more general discussion see Blumenthal and Getoor [1], Ch. III.

<u>C) The martingale problem</u>

If $dX_t = b(X_t)dt + \sigma(X_t)dB_t$ is an Ito diffusion in $\mathbb{R}^n$ with generator A and if $f \in C_0^2(\mathbb{R}^n)$ then by (7.20)

$$(8.11) \qquad f(X_t) = f(x) + \int_0^t Af(X_s)ds + \int_0^t \nabla f^T(X_s)\sigma(X_s)dB_s$$

Define

$$(8.12) \qquad M_t = f(X_t) - \int_0^t Af(X_r)dr \quad \left(= f(x) + \int_0^t \nabla f^T(X_r)\sigma(X_r)dB_r\right)$$

Then, since Ito integrals are martingales (wrt. the σ-algebras $\{\mathcal{F}_t\}$) we have for $s > t$

$$E^x[M_s | \mathcal{F}_t] = M_t$$

It follows that

$$E^x[M_s | \mathcal{F}_t] = E^x[E^x[M_s | \mathcal{F}_t] | \mathcal{M}_t] = E^x[M_t | \mathcal{M}_t] = M_t,$$

since M_t is $\mathcal{M}_t$-measurable. We have proved:

THEOREM 8.7. If X_t is an Ito diffusion in $\mathbb{R}^n$ with generator A, then for all $f \in C_0^2(\mathbb{R}^n)$ the process

$$M_t = f(X_t) - \int_0^t Af(X_r)dr$$

is a martingale wrt. $\{\mathcal{M}_t\}$.

If we identify each $\omega \in \Omega$ with the function

$$\omega_t = \omega(t) = X_t^x(\omega)$$

we see that the probability space $(\Omega, \mathcal{M}, Q^x)$ is identified with

$$((\mathbb{R}^n)^{[0,\infty)}, \mathcal{B}, \tilde{Q}^x)$$

where $\mathcal{B}$ is the Borel σ-algebra on $(\mathbb{R}^n)^{[0,\infty)}$ (see Ch.II). Thus, regarding the law of X_t^x as a probability measure $\tilde{Q}^x$ on $\mathcal{B}$ we can formulate Theorem 8.7 as follows:

THEOREM 8.7'. If $\tilde{Q}^x$ is the probability measure on $\mathcal{B}$ induced by the law Q^x of an Ito diffusion X_t, then for all $f \in C_0^2(\mathbb{R}^n)$ the process

$$(8.13) \qquad M_t = f(X_t) - \int_0^t Af(X_r)dr \; (= f(\omega_t) - \int_0^t Af(\omega_r)dr); \; \omega \in (\mathbb{R}^n)^{[0,\infty)}$$

is a $\tilde{Q}^x$-maringale wrt. the Borel σ-algebras $\mathcal{B}_t$ of $(\mathbb{R}^n)^{[0,t]}$, $t \geq 0$. In other words, the measure $\tilde{Q}^x$ solves the martingale problem for the differential operator A, in the following sense:

DEFINITION 8.8. Let L be a semi-elliptic differential operator of the form

$$L = \Sigma b_i \frac{\partial}{\partial x_i} + \sum_{i,j} a_{ij} \frac{\partial^2}{\partial x_i \partial x_j}$$

where the coefficients b_i, a_{ij} are locally bounded Borel measurable functions on $\mathbb{R}^n$. Then we say that a probability measure $\tilde{P}^x$ on $((\mathbb{R}^n)^{[0,\infty)}, \mathcal{B})$ solves the martingale problem for L (starting at x) if the process

$$M_t = f(\omega_t) - \int_0^t Lf(\omega_r)dr, \quad M_0 = f(x) \quad \text{a.s. } \tilde{P}^x$$

is a $\tilde{P}^x$ martingale wrt. $\mathcal{B}_t$, for all $f \in C_0^2(\mathbb{R}^n)$. The martingale problem is called well posed if there is a unique measure $\tilde{P}^x$ solving the martingale problem.

The argument of Theorem 8.7 actually proves that $\tilde{Q}^x$ solves the martingale problem for A whenever X_t is a weak solution of the stochastic differential equation

(8.14) $$dX_t = b(X_t)dt + \sigma(X_t)dB_t$$

Conversely, it can be proved that if $\tilde{P}^x$ solves the martingale problem for

(8.15) $$L = \Sigma b_i \frac{\partial}{\partial x_i} + \frac{1}{2}\Sigma(\sigma\sigma^T)_{ij}\frac{\partial^2}{\partial x_i \partial x_j}$$

starting at x, for all $x \in \mathbb{R}^n$, then there exists a weak solution X_t of the stochastic differential equation (8.14). Moreover, this weak solution X_t is a Markov process if an only if the martingale problem for L is well posed. (See Stroock and Varadhan [1] or Rogers and Williams [1]). Therefore, if the coefficients b, σ of (8.14) satisfy the conditions (5.14), (5.15) of Theorem 5.5, we conclude that

(8.16) $\tilde{Q}^x$ is the <u>unique</u> solution of the martingale problem for the operator L given by (8.15).

Lipschitz-continuity of the coefficients of L is not <u>necessary</u> for the uniqueness of the martingale problem. For example, one of the spectacular results of Stroock and Varadhan [1] is that

$$L = \Sigma b_i \frac{\partial}{\partial x_i} + \Sigma a_{ij}\frac{\partial^2}{\partial x_i \partial x_j}$$

has a unique solution of the martingale problem if $[a_{ij}]$ is everywhere positive definite, $a_{ij}(x)$ is continuous and there exists a constant D such that

$$|b(x)| + |a(x)| \leq D(1+|x|) \quad \text{for all} \quad x \in \mathbb{R}^n.$$

<u>D) When is a stochastic integral a diffusion?</u>

The Ito formula gives that if we apply a C^2 function $\phi: U \subset \mathbb{R}^n \to \mathbb{R}^n$ to a stochastic integral X_t the result $\phi(X_t)$ is another stochastic integral. A natural question is: If X_t is an Ito diffusion will $\phi(X_t)$ be an Ito diffusion too? The answer to this question is no in general, but it may be yes in some cases:

<u>EXAMPLE 8.9</u>. (The Bessel process). In Example 5.3 we found that the process

$$R = \sqrt{B_1^2 + B_2^2}$$

satisfied the equation

(8.17) $$dR = B_1(B_1^2+B_2^2)^{-\frac{1}{2}}dB_1 + B_2(B_1^2+B_2^2)^{-\frac{1}{2}}dB_2 + \tfrac{1}{2}(B_1^2+B_2^2)^{-\frac{1}{2}}dt$$

However, as it stands this is not a stochastic differential equation of the form (5.17), so it is not apparent from (8.17) that R is an Ito diffusion. But this will follow if we can show that

$$Y_t = \int_0^t B_1(B_1^2+B_2^2)^{-\frac{1}{2}}dB_1 + \int_0^t B_2(B_1^2+B_2^2)^{-\frac{1}{2}}dB_2$$

coincides in law with (i.e. has the same finite-dimensional distributions as) 1-dimensional Brownian motion $\tilde{B}_t$. For then (8.17) can be written

$$dR = \frac{1}{2R}\,dt + d\tilde{B}$$

which is of the form (5.17), thus showing by weak uniqueness (5.28) that R is an Ito diffusion with charateristic operator

$$f(x) = \tfrac{1}{2}f''(x) + \frac{1}{2x}\,f'(x)$$

as claimed in Example 4.7. One way of seeing that the process Y_t coincides in law with 1-dimensional Brownian motion $\tilde{B}_t$ is to apply the following result:

THEOREM 8.10. A stochastic integral

$$dY_t = v\,dB,\quad Y_0 = 0 \quad \text{with} \quad v(t,\omega) \in \mathbb{R}^{n\times m}$$

coincides (in law) with n-dimensional Brownian motion if and only if

(8.18) $$vv^T(t,\omega) = I_n \quad \text{for a.a. } (t,\omega) \text{ wrt. } dt \times dP^0,$$

where I_n is the n-dimensional identity matrix and P^0 is the probability law of m-dimensional Brownian motion starting at 0.

Note that in the example above we have

$$Y_t = \int_0^t v\,dB$$

with

$$v = [B_1(B_1^2+B_2^2)^{-\frac{1}{2}},\ B_2(B_1^2+B_2^2)^{-\frac{1}{2}}],\quad B = \begin{bmatrix} B_1 \\ B_2 \end{bmatrix}$$

and since $vv^T = 1$, we get that Y_t is 1-dimensional Brownian motion, as required.

Theorem 8.10 is a special case of the following result, which gives a necessary and sufficient condition for a stochastic integral to coincide in law with a given diffusion: (We use the symbol $\sim$ for "coincides in law with")

THEOREM 8.11. Let X_t be an Ito diffusion

$$dX_t = b(X_t)dt + \sigma(X_t)dB_t, \quad b \in \mathbb{R}^n, \ \sigma \in \mathbb{R}^{n\times m}, \quad X_0 = x,$$

and let Y_t be a stochastic integral

$$dY_t = u(t,\omega)dt + v(t,\omega)dB_t, \quad u \in \mathbb{R}^n, \quad v \in \mathbb{R}^{n\times m}, \quad Y_0 = x.$$

Then $X_t \sim Y_t$ if and only if

$$E^x[u(t,\cdot)|\mathcal{N}_t] = b(Y_t^x) \quad \text{and} \quad vv^T(t,\omega) = \sigma\sigma^T(Y_t^x) \quad \text{for all } t,\omega, \tag{8.19}$$

for a.a. (t,ω) wrt. $m\times P^0$, where $\mathcal{N}_t$ is the σ-algebra generated by Y_s; $s\leq t$, and m is Lebesgue measure on $\mathbb{R}$.

Proof. Assume that (8.19) holds. Let

$$A = \Sigma b_i \frac{\partial}{\partial x_i} + \frac{1}{2} \sum_{i,j} (\sigma\sigma^T)_{ij} \frac{\partial^2}{\partial x_i \partial x_j}$$

be the generator of X_t and define, for $f \in C_0^2(\mathbb{R}^n)$,

$$Hf(t,\omega) = \sum_i u_i(t,\omega) \frac{\partial f}{\partial x_i}(Y_t) + \frac{1}{2} \sum_{i,j} (vv^T)_{ij}(t,\omega) \frac{\partial^2 f}{\partial x_i \partial x_j}(Y_t)$$

Then by Ito's formula (see (7.20)) we have, for $s>t$,

$$E^x[f(Y_s)|\mathcal{N}_t] = f(Y_t) + E^x[\int_t^s Hf(r,\omega)dr|\mathcal{N}_t] + E^x[\int_t^s \nabla f^T v dB_r|\mathcal{N}_t] \tag{8.20}$$

$$= f(Y_t) + E^x[\int_t^s E^x[Hf(r,\omega)|\mathcal{N}_r]dr|\mathcal{N}_t] + 0$$

$$= f(Y_t) + E^x[\int_t^s Af(Y_r)dr \ |\mathcal{N}_t], \quad \text{by (8.19)}$$

where E^x denotes expectation wrt. the law R^x of Y_t (see Lemma 7.8). Therefore, if we define

$$M_t = f(Y_t) - \int_0^t Af(Y_r)dr \tag{8.21}$$

then, for $s>t$,

$$E^x[M_s|\mathcal{N}_t] = f(Y_t) + E^x[\int_t^s Af(Y_r)dr|\mathcal{N}_t] - E^x[\int_0^s Af(Y_r)dr|\mathcal{N}_t]$$

$$= f(Y_t) - E^x[\int_0^t Af(Y_r)dr|\mathcal{N}_t] = M_t$$

Hence M_t is a martingale wrt. the σ-algebras $\mathcal{N}_t$ and the law R^x. By uniqueness of the solution of the martingale problem (see (8.16)) we conclude that $X_t \sim Y_t$.

Conversely, assume that $X_t \sim Y_t$. Choose $f \in C_0^2$. By Ito's formula (7.20) we have, for a.a.(t,ω) wrt. $m \times P^0$,

$$(8.22) \qquad \lim_{h \downarrow 0} \frac{1}{h} (E^x[f(Y_{t+h}) | \mathcal{N}_t] - f(Y_t))$$

$$= \lim_{h \downarrow 0} \frac{1}{h} \Big(\int_t^{t+h} E^x[\sum_i u_i(s,\omega) \frac{\partial f}{\partial x_i}(Y_s) + \tfrac{1}{2} \sum_{i,j} (vv^T)_{ij}(s,\omega) \frac{\partial^2 f}{\partial x_i \partial x_j}(Y_s) | \mathcal{N}_t] ds \Big)$$

$$(8.23) \quad = \sum_i E^x[u_i(t,\omega) | \mathcal{N}_t] \frac{\partial f}{\partial x_i}(Y_t) + \tfrac{1}{2} \sum_{i,j} E^x[(vv^T)_{ij}(t,\omega) | \mathcal{N}_t] \frac{\partial^2 f}{\partial x_i \partial x_j}(Y_t)$$

On the other hand, since $X_t \sim Y_t$ we know that Y_t is a Markov process Therefore (8.22) coincides with

$$\lim_{h \downarrow 0} \frac{1}{h} (E^{Y_t}[f(Y_h)] - E^{Y_t}[f(Y_0)])$$

$$= \sum_i E^{Y_t}[u_i(0,\omega) \frac{\partial f}{\partial x_i}(Y_0)] + \tfrac{1}{2} \sum_{i,j} E^{Y_t}[(vv^T)_{ij}(0,\omega) \frac{\partial^2 f}{\partial x_i \partial x_j}(Y_0)]$$

$$(8.24) \qquad = \sum_i E^{Y_t}[u_i(0,\omega)] \frac{\partial f}{\partial x_i}(Y_t) + \tfrac{1}{2} \sum_{i,j} E^{Y_t}[(vv^T)_{ij}(0,\omega)] \frac{\partial^2 f}{\partial x_i \partial x_j}(Y_t)$$

Comparing (8.23) and (8.24) we conclude that

$$(8.25) \qquad E^x[u(t,\omega) | \mathcal{N}_t] = E^{Y_t}[u(0,\omega)] \text{ and } E^x[vv^T(t,\omega) | \mathcal{N}_t] = E^{Y_t}[vv^T(0,\omega)]$$

for a.a.(t,ω).

On the other hand, since the generator of Y_t coincides with the generator A of X_t we get from (8.24) that

$$(8.26) \qquad E^{Y_t}[u(0,\omega)] = b(Y_t) \quad \text{and} \quad E^{Y_t}[vv^T(0,\omega)] = \sigma\sigma^T(Y_t)$$

for a.a.(t,ω).

Combining (8.25) and (8.26) we conclude that

$$(8.27) \qquad E^x[u | \mathcal{N}_t] = b(Y_t) \quad \text{and} \quad E^x[vv^T | \mathcal{N}_t] = \sigma\sigma^T(Y_t) \text{ for a.a.}(t,\omega).$$

From this we obtain (8.19) by using that in fact $vv^T(t,\cdot)$ is always $\mathcal{N}_t$-measurable, in the following sense:

<u>LEMMA 8.12</u>. Let $dY_t = u(t,\omega)dt + v(t,\omega)dB_t$, $Y_0 = x$ be as in Theorem 8.11. Then there exists an $\mathcal{N}_t$-adapted process $W(t,\omega)$ such

that

$$vv^T(t,\omega) = W(t,\omega) \quad \text{for} \quad \text{a.a.}(t,\omega)$$

Proof. By Ito's formula we have (if $Y_i(t,\omega)$ denotes component number i of $Y(t,\omega)$)

$$Y_iY_j(t,\omega) = x_ix_j + \int_0^t Y_i dY_j(s) + \int_0^t Y_j dY_i(s) + \int_0^t (vv^T)_{ij}(s,\omega)ds$$

Therefore, if we put

$$H_{ij}(t,\omega) = Y_iY_j(t,\omega) - x_jx_i - \int_0^t Y_i dY_j - \int_0^t Y_j dY_i, \quad 1 \le i,j \le n$$

then H_{ij} is $\mathcal{N}_t$-adapted and

$$H_{ij}(t,\omega) = \int_0^t (vv^T)_{ij}(s,\omega)ds$$

Therefore

$$(vv^T)_{ij}(t,\omega) = \lim_{r \downarrow 0} \frac{H(t,\omega) - H(t-r,\omega)}{r}$$

for a.a. t. This shows Lemma 8.12 and the proof of Theorem 8.11 is complete.

REMARKS. 1) One may ask if it also true that $u(t,\cdot)$ must be $\mathcal{N}_t$-measurable. However, the following example shows that this fails even in the case when $v=n=1$:

Let B_1, B_2 be two independent 1-dimensional Brownian motions and define

$$dY_t = B_1(t)dt + dB_2(t)$$

Then we may regard Y_t as noisy observations of the process $B_1(t)$. So by Example 6.11 we have that

$$E^0[(B_1(t,\omega) - \hat{B}_1(t,\omega))^2] = \tanh(t),$$

where $\hat{B}_1(t,\omega) = E^0[B_1(t)|\mathcal{N}_t]$ is the Kalman-Bucy filter. In particular, $B_1(t,\omega)$ cannot be $\mathcal{N}_t$-measurable.

2) The process $v(t,\omega)$ need not be $\mathcal{N}_t$-adapted either: Let B_t be 1-dimensional Brownian motion and define

$$(8.28) \qquad dY_t = \text{sign}(B_t)dB_t$$

where

$$\text{sign}(z) = \begin{cases} 1 & \text{if } z > 0 \\ -1 & \text{if } z < 0 \end{cases}$$

Tanaka's formula says that

(8.29) $$|B_t| = |B_0| + \int_0^t \text{sign}(B_s)dB_s + L_t$$

where $L_t = L_t(\omega)$ is local time of B_t at 0, a non-decreasing process which only increases when $B_t = 0$. (see e.g. Chung and Williams [1]). Therefore the σ-algebra $\mathcal{N}_t$ generated by $\{Y_s; s\leq t\}$ is contained in the σ-algebra $\mathcal{H}_t$ generated by $\{|B_s| ; s\leq t\}$. It follows that $v(t,\omega) = \text{sign}(B_t)$ cannot be $\mathcal{N}_t$-adapted.

COROLLARY 8.13 (How to recognize a Brownian motion).
Let

$$dY_t = u(t,\omega)dt + v(t,\omega)dB_t$$

be a stochastic integral in $\mathbb{R}^n$. Then Y_t is a Brownian motion if and only if

(8.30) $$E^x[u(t,\cdot) | \mathcal{N}_t] = 0 \quad \text{and} \quad vv^T(t,\omega) = I_n$$

for a.a.(t,ω).

REMARK.
Using Theorem 8.11 one may now proceed to investigate when the image $Y_t = \phi(X_t)$ of an Ito diffusion X_t by a C^2-function ϕ coincide in law with an Ito diffusion Z_t. Applying the criterion (8.19) one obtains the following result:

(8.31) $$\phi(X_t) \sim Z_t \quad \text{if and only if}$$

$$A[f\circ\phi] = \hat{A}[f] \circ \phi$$

for all second order polynomials $f(x_1,\dots,x_n) = \Sigma a_i x_i + \Sigma c_{ij} x_i x_j$, $f(x_1,\dots,x_n)$ (and hence for all $f \in C_0^2$) where A and $\hat{A}$ are the generators of X_t and Z_t respectively. For generalizations of this result, see Csink and Øksendal [1], [2] and Øksendal [1-3].

E) Random time change
Let $c(t,\omega) \geq 0$ be an $\mathcal{F}_t$-adapted process. Define

(8.32) $$\beta_t = \beta(t,\omega) = \int_0^t c(s,\omega)ds$$

We will say that β_t is a (random) time change with time change rate $c(t,\omega)$.

Note that $\beta(t,\omega)$ is also $\mathcal{F}_t$-adapted and for each ω the map $t\to\beta_t(\omega)$ is non-decreasing. Let $\alpha_t=\alpha(t,\omega)$ be the right-continuous inverse of β_t:

(8.33) $\quad \alpha_t = \inf\{s;\ \beta_s > t\}$

Then $\omega\to\alpha(t,\omega)$ is an $\{\mathcal{F}_s\}$-stopping time for each t, since $\{\omega;\ \alpha(t,\omega)<s\} = \{\omega; t<\beta(s,\omega)\} \in \mathcal{F}_s$.

We now ask the question: Suppose X_t is an Ito diffusion and Y_t a stochastic integral as in Theorem 8.11. When does there exist a time change β_t sucht that $Y_{\alpha_t} \sim X_t$? (Note that α_t is only defined up to time β_∞. If $\beta_\infty < \infty$ we interpret $Y_{\alpha_t} \sim X_t$ to mean that Y_{α_t} has the same law as X_t up to time β_∞).

Basically the same method of proof as in Theorem 8.11 extends to give the following partial answer (see Øksendal [3]).

<u>THEOREM 8.14</u>. Let X_t, Y_t be as in Theorem 8.11 (except that the continuity conditions on u and vv^T are not needed) and let β_t be a time change as above. Assume that

(8.35) $\quad u(t,\omega) = c(t,\omega)b(Y_t) \quad$ and $\quad vv^T(t,\omega) = c(t,\omega)\cdot\sigma\sigma^T(t,\omega)$

for a.a. t,ω. Then

$$Y_{\alpha_t} \sim X_t$$

This result allows us to recognize time changes of Brownian motion:

<u>THEOREM 8.15</u>. Let $dY_t = v(t,\omega)dB_t$, $v \in \mathbb{R}^{n\times m}$, $B_t \in \mathbb{R}^m$ be an Ito integral in $\mathbb{R}^n$, $Y_0 = 0$ and assume that

(8.36) $\quad vv^T(t,\omega) = c(t,\omega)I_n$

for some process $c(t,\omega)$. Then

Y_{α_t} is n-dimensional Brownian motion.

(Here - and in the following - α,β and c are related as in (8.32), (8.33))

<u>COROLLARY 8.16</u>. Let $dY_t = \sum_{i=1}^{n} v_i(t,\omega)dB_i(t,\omega)$, $Y_o = 0$, where $B=(B_1,\ldots,B_n)$ is Brownian motion in $\mathbb{R}^n$. Then

Y_{α_t} is 1-dimensional Brownian motion,

where $\beta_s = \int_0^s \{ \sum_{i=1}^n v_i^2(r,\omega)\} dr$.

COROLLARY 8.17. Let $c(t,\omega)$ be given and define

$$dY_t = \int_0^t \sqrt{c(s,\omega)}\, dB_s,$$

where B_s is n-dimensional Brownian motion. Then

Y_{α_t} is also an n-dimensional Brownian motion.

We now use this to prove that a time change of an Ito integral is again an Ito integral, but driven by a different Brownian motion $\tilde{B}_t$. First we construct $\tilde{B}_t$:

LEMMA 8.18. Suppose $s \to \alpha(s,\omega)$ is continuous, $\alpha(0,\omega)=0$ for a.a.ω. Fix $t>0$ such that $\beta_t < \infty$ a.s. and assume that $E[\alpha_t] < \infty$. For $k=1,2,\ldots$ put

$$t_j = \begin{cases} j \cdot 2^{-k} & \text{if} \quad j \cdot 2^{-k} \leq t \\ t & \text{if} \quad j \cdot 2^{-k} > t \end{cases}$$

and choose r_j such that $\alpha_{r_j} = t_j$. Suppose $f(s,\omega) \geq 0$ is $\mathcal{F}_s$-adapted bounded and s-continuous for a.a.ω. Then

$$\lim_{k\to\infty} \sum_j f(\alpha_j,\omega)\Delta B_{\alpha_j} = \int_0^{\alpha_t} f(s,\omega) dB_s \quad \text{a.s.,} \tag{8.37}$$

where $\alpha_j = \alpha_{r_j}$, $\Delta B_{\alpha_j} = B_{\alpha_{j+1}} - B_{\alpha_j}$ and the limit is in $L^2(\Omega, P^0)$.

Proof. For all k we have

$$E[(\sum_j f(\alpha_j,\omega)\Delta B_{\alpha_j} - \int_0^{\alpha_t} f(s,\omega)dB_s)^2]$$

$$= E[(\sum_j \int_{\alpha_j}^{\alpha_{j+1}} (f(\alpha_j,\omega)-f(s,\omega)dB_s)^2] = \sum_j E[(\int_{\alpha_j}^{\alpha_{j+1}} (f(\alpha_j,\omega)-f(s,\omega))dB_s)^2]$$

$$= \sum_j E[(\int_{\alpha_j}^{\alpha_{j+1}} (f(\alpha_j,\omega)-f(s,\omega))^2 ds] = E[(\int_0^{\alpha_t} (f-f_k)^2 ds],$$

where $f_k(s,\omega) = \sum_j f(t_j,\omega)\chi_{[t_j,t_{j+1})}(s)$ is the elementary approximation to f. (See Ch. III). This implies (8.37).

We now use this to establish a general time change formula for Ito integrals. An alternative proof in the case $n=m=1$ can be found in McKean [1, §2.8].

THEOREM 8.19. (Time change formula for Ito integrals). Suppose $c(s,\omega)$ and $\alpha(s,\omega)$ are s-continuous, $\alpha(0,\omega)=0$ for a.a.ω and that $E[\alpha_t]<\infty$. Let $(B_s,\mathcal{F}_s)$ be m-dimensional Brownian motion and let $v(s,\omega)\in\mathbb{R}^{n\times m}$ be bounded and s-continuous. Define

$$(8.38)\qquad \tilde{B}_t=\lim_{k\to\infty}\sum_j\sqrt{c(\alpha_j,\omega)}\,\Delta B_{\alpha_j}=\int_0^{\alpha_t}\sqrt{c(s,\omega)}\,dB_s$$

Then $\tilde{B}_t$ is an (m-dimensional) $\mathcal{F}_{\alpha_t}$-Brownian motion (i.e. $\tilde{B}_t$ is a Brownian motion and $\tilde{B}_t$ is a martingale wrt. $\mathcal{F}_{\alpha_t}$) and

$$(8.39)\qquad \int_0^{\alpha_t}v(s,\omega)dB_s=\int_0^t v(\alpha_r,\omega)\sqrt{\alpha_r'(\omega)}\,d\tilde{B}_r,\quad \text{a.s. } P^0,$$

where $\alpha_r'(\omega)$ is the derivative of $\alpha(r,\omega)$ wrt. r, so that

$$(8.40)\qquad \alpha_r'(\omega)=\frac{1}{c(\alpha_r,\omega)}\qquad \text{for a.a. } r\geq 0,\ \text{a.a. } \omega\in\Omega.$$

Proof. The existence of the limit in (8.38) and the second identity in (8.38) follows by applying Lemma 8.18 to the function

$$f(s,\omega)=\sqrt{c(s,\omega)}.$$

Then by Corollary 8.17 we have that $\tilde{B}_t$ is an $\mathcal{F}_{\alpha_t}$-Brownian motion. It remains to prove (8.39):

$$\begin{aligned}\int_0^{\alpha_t}v(s,\omega)dB_s&=\lim_{k\to\infty}\sum_j v(\alpha_j,\omega)\Delta B_{\alpha_j}\\&=\lim_{k\to\infty}\sum_j v(\alpha_j,\omega)\sqrt{\frac{1}{c(\alpha_j,\omega)}}\cdot\sqrt{c(\alpha_j,\omega)}\Delta B_{\alpha_j}\\&=\lim_{k\to\infty}\sum_j v(\alpha_j,\omega)\sqrt{\frac{1}{c(\alpha_j,\omega)}}\,\Delta\tilde{B}_j\\&=\int_0^t v(\alpha_r,\omega)\sqrt{\frac{1}{c(\alpha_r,\omega)}}\,d\tilde{B}_r\end{aligned}$$

and the proof is complete.

EXAMPLE 8.20. (Brownian motion on the unit sphere in $\mathbb{R}^n$; $n>2$). In Examples 5.4 and 7.18 we constructed Brownian motion on the unit circle. It is not obvious how to extend the method used there to obtain Brownian motion on the unit sphere S of $\mathbb{R}^n$; $n\geq 3$. However, we may proceed as follows: Apply the function $\phi:\mathbb{R}^n\setminus\{0\}\to S$ defined by

$$\phi(x) = x \cdot |x|^{-1};\ x \in \mathbb{R}^n \setminus \{0\}$$

to n-dimensional Brownian motion $B = (B_1,\ldots,B_n)$. The result is a stochastic integral $Y = (Y_1,\ldots,Y_n) = \phi(B)$ which by Ito's formula is given by

$$(8.41)\qquad dY_i = \frac{|B|^2 - B_i^2}{|B|^3}\, dB_i - \sum_{j\neq i} \frac{B_j B_i}{|B|^3}\, dB_j - \frac{n-1}{2} \cdot \frac{B_i}{|B|^2}\, dt;\ i=1,2,\ldots,n$$

Hence

$$dY = \frac{1}{|B|} \cdot \sigma(Y)dB + \frac{1}{|B|^2}\, b(Y)dt,$$

where

$$\sigma = [\sigma_{ij}] \in \mathbb{R}^{n\times m},\quad \text{with}\quad \sigma_{ij}(Y) = \delta_{ij} - Y_i Y_j;\ 1 \leq i,\ j \leq n$$

and

$$b(y) = -\frac{n-1}{2} \cdot \begin{bmatrix} y_1 \\ \vdots \\ y_n \end{bmatrix} \in \mathbb{R}^n,\ (y_1,\ldots,y_n \text{ are the coordinates of } y \in \mathbb{R}^n).$$

Now perform the following time change: Define

$$Z_t(\omega) = Y_{\alpha(t,\omega)}(\omega)$$

where

$$\alpha_t = \beta_t^{-1},\ \beta(t,\omega) = \int_0^t \frac{1}{|B|^2}\, ds.$$

Then Z is again a stochastic integral and by Theorem 8.19

$$dZ = \sigma(Z)dB + b(Z)dt.$$

Hence Z is a diffusion with characteristic operator

$$(8.42)\qquad \mathscr{A}f(y) = \tfrac{1}{2}\Big(\Delta f(y) - \sum_{i,j} y_i y_j \frac{\partial^2 f}{\partial y_i \partial y_j}\Big) - \frac{n-1}{2} \cdot \sum_i y_i \frac{\partial f}{\partial y_i};\ |y|=1$$

Thus, $\phi(B) = \frac{B}{|B|}$ is - after a suitable change of time scale - equal to a diffusion Z living on the unit sphere S of $\mathbb{R}^n$. (This result could also have been obtained by using Theorem 1 in Csink and Øksendal [1]). Note that Z is invariant under orthogonal transformations in $\mathbb{R}^n$ (since B is). It is reasonable to call Z <u>Brownian motion on the unit sphere S</u>. For other constructions see Ito and McKean [1, p. 269 (§7.15)] and Stroock [1].

More generally, given a Riemannian manifold M with metric tensor $g = [g_{ij}]$ one may define a <u>Brownian motion on M</u> as a diffusion on M whose characteristic operator $\mathscr{A}$ in local coordinates x_i is given by ½ times the Laplace-Beltrami operator (here $[g^{ij}]=[g_{ij}]^{-1}$)

$$(8.43)\qquad \Delta_M = \frac{1}{\sqrt{\det(g)}} \cdot \sum_i \frac{\partial}{\partial x_i} \left(\sqrt{\det(g)} \sum_j g^{ij} \frac{\partial}{\partial x_j} \right)$$

See for example Meyer [1, p. 256-270], McKean [1, §4.3]. The subject of stochastic differential equatons on manifolds is also treated in Ikeda and Watanabe [1] and Elworthy [1].

EXAMPLE 8.21. (Harmonic and analytic functions)
Let $B = (B_1, B_2)$ be 2-dimensional Brownian motion. Let us investigate what happens if we apply a C^2 function

$$\phi(x_1,x_2) = (u(x_1,x_2),\ v(x_1,x_2))$$

to B:
Put $Y = (Y_1,Y_2) = \phi(B_1,B_2)$ and apply Ito's formula:

$$dY_1 = u_1'(B_1,B_2)dB_1 + u_2'(B_1,B_2)dB_2 + \tfrac{1}{2}[u_{11}''(B_1,B_2) + u_{22}''(B_1,B_2)]dt$$

and

$$dY_2 = v_1'(B_1,B_2)dB_1 + v_2'(B_1,B_2)dB_2 + \tfrac{1}{2}[v_{11}''(B_1,B_2) + v_{22}''(B_1,B_2)]dt,$$

where $u_1' = \frac{\partial u}{\partial x_1}$ etc.
So

$$dY = b(B_1,B_2)dt + \sigma(B_1,B_2)dB,$$

with $b = \tfrac{1}{2}\begin{bmatrix}\Delta u\\ \Delta v\end{bmatrix}$, $\sigma = \begin{bmatrix} u_1' & u_2' \\ v_1' & v_2' \end{bmatrix} = D_\phi$ (the derivative of ϕ)

So $Y = \phi(B_1,B_2)$ is a martingale if (and, in fact, only if) ϕ is harmonic, i.e. $\Delta\phi = 0$.
If ϕ is harmonic, we get by Corollary 8.16 that

$$\phi(B_1,B_2) = (\tilde{B}^{(1)}_{\beta_1},\ \tilde{B}^{(2)}_{\beta_2})$$

where $\tilde{B}^{(1)}$ and $\tilde{B}^{(2)}$ are two (not necessarily independent) versions of 1-dimensional Brownian motion, and

$$\beta_1(t,\omega) = \int_0^t |\nabla u|^2(B_1,B_2)ds,\quad \beta_2(t,\omega) = \int_0^t |\nabla v|^2(B_1,B_2)ds.$$

Since

$$\sigma\sigma^T = \begin{bmatrix} |\nabla u|^2 & \nabla u\cdot\nabla v \\ \nabla u\cdot\nabla v & |\nabla v|^2 \end{bmatrix}$$

we see that if (in addition to $\Delta u = \Delta v = 0$)

$$(8.43)\qquad |\nabla u|^2 = |\nabla v|^2 \quad\text{and}\quad \nabla u \cdot \nabla v = 0$$

then

$$Y_t = Y_0 + \int_0^t \sigma\, dB$$

with

$$\sigma\sigma^T = |\nabla u|^2(B_1,B_2)I_2, \quad Y_0 = \phi(B_1(0), B_2(0)).$$

Therefore, if we let

$$\beta_t = \beta(t,\omega) = \int_0^t |\nabla u|^2(B_1,B_2)ds, \quad \alpha_t = \beta_t^{-1}$$

we obtain by Theorem 8.15 that Y_{α_t} is <u>2-dimensional Brownian motion</u>. Conditions (8.43) - in addition to $\Delta u = \Delta v = 0$ - are easily seen to be equivalent to requiring that the function $\phi(x+iy)=\phi(x,y)$ regarded as a complex function is either <u>analytic</u> or <u>conjugate analytic</u>.

Thus we have proved a theorem of P. Lévy that $\phi(B_1,B_2)$ is - after a change of time scale - again Brownian motion in the plane if and only if ϕ is either analytic or conjugate analytic. For extensions of this result see Bernard, Campbell and Davie [1], Csink and Øksendal [1-2] and Øksendal [1-3].

<u>F) The Cameron-Martin-Girsanov formula</u>

We end this chapter by proving a result which is useful for example in connection with questions of existence of diffusions with a given infinitesimal generator where the drift coefficients are not necessarily continuous. This question is important in stochastic control theory (see Chapter XI). For more information see for example Ikeda and Watanabe [1] or Stroock and Varadhan [1].

<u>THEOREM 8.22</u>. (The Cameron-Martin-Girsanov formula I)
Let X_t be an Ito diffusion given by

(8.44) $\quad dX_t = b(X_t)dt + \sigma(X_t)dB_t, \; X_0 = x, \; b \in \mathbb{R}^n, \; \sigma \in \mathbb{R}^{n\times n}$

and assume that $\sigma(x)$ is invertible for all x and that $\sigma^{-1}(x)$ is bounded. Suppose Y_t satisfies the equation

(8.45) $\quad dY_t = a(t,\omega)dt + b(Y_t)dt + \sigma(Y_t)dB_t, \; Y_0 = x$

where $a(t,\omega) \in N(0,\infty)$ (see Definition 3.4) and let Z_t be the (real-valued) stochastic integral

(8.46) $\quad dZ_t = -\tfrac{1}{2}(\sigma^{-1}(Y_t)a(t,\omega))^2dt - (\sigma^{-1}(Y_t)a(t,\omega))^T dB_t, \; Z_0 = 0$

and put

(8.47) $$M_t = e^{Z_t}$$

Then for all $0 \leq t_1 \leq \ldots \leq t_k \leq t$ and $f_i \in C_0^2$ we have

(8.48) $$E^x[M_t f_1(Y_{t_1}) \ldots f_k(Y_{t_k})] = E^x[f_1(X_{t_1}) \ldots f_k(X_{t_k})].$$

REMARK. If we regard the laws Q^x, R^x of X_t^x, Y_t^x as probability measures $\tilde{Q}^x$, $\tilde{R}^x$ on $((\mathbb{R}^n)^{[0,\infty)}, \mathscr{B})$ via the identifications $\omega(\cdot) \leftrightarrow Y^x(\omega)$, $\omega(\cdot) \leftrightarrow X^x(\omega)$ (as done before Theorem 8.7') then (8.48) may be stated as follows:

(8.49) The law of $\{X_s\}_{s \leq t}$ (i.e. the measure $\tilde{Q}^x$ restricted to $\mathscr{B}_t$, the family of Borel subsets of $(\mathbb{R}^n)^{[0,t]}$) is absolutely continuous wrt. the law of $\{Y_s\}_{s \leq t}$, with Radon-Nikodym derivative M_t.

Proof. First note that by Ito's formula we have

$$dM_t = e^{Z_t} dZ_t + \tfrac{1}{2} e^{Z_t} (dZ_t)^2 = -M_t (\sigma^{-1} a)^T dB_t.$$

Therefore M_t is an $\mathscr{F}_t$-martingale, so if $t > s$ we have

$$E[M_t | \mathcal{N}_s] = E[E[M_t | \mathscr{F}_s] | \mathcal{N}_s] = E[M_s | \mathcal{N}_s],$$

where $\mathcal{N}_s$ is the σ-algebra generated by $\{Y_u;\ u \leq s\}$. Therefore, if $f \in C_0^2$ and $t \geq s > r$ we have

(8.50) $$E^x[M_t f(Y_s) | \mathcal{N}_r] = E^x[E^x[M_t f(Y_s) | \mathcal{N}_s] | \mathcal{N}_r] = E^x[M_s f(Y_s) | \mathcal{N}_r]$$

Choose $f \in C_0^2$ and consider

$$K_t = M_t S_t, \quad \text{where} \quad S_t = f(Y_t).$$

By Ito's formula and (7.20) we have

$$dK_t = M_t dS_t + S_t dM_t + (dM_t)(dS_t)$$

$$= M_t[a^T(t,\omega) \nabla f(Y_t) dt + (Af)(Y_t) dt - ((\sigma^{-1}(Y_t) a(t,\omega))^T dB_t) \cdot$$

$$\cdot (\nabla f^T(Y_t) \sigma(Y_t) dB_t)] + dB_t\text{-terms}$$

$$= M_t[a^T(t,\omega) \nabla f(Y_t) dt + (Af)(Y_t) dt - a^T(t,\omega) \nabla f(Y_t) dt] + dB_t\text{-terms}$$

$$= M_t (Af)(Y_t) dt + dB_t\text{-terms},$$

where A is the generator of X_t. So if $s > r$ then

$$(8.51)\qquad E^x[M_s f(Y_s)|\mathcal{N}_r] = f(Y_r)E[M_r|\mathcal{N}_r] + E^x[\int_r^s M_u\cdot Af(Y_u)du|\mathcal{N}_r]$$

Define

$$(8.52)\qquad J_s = f(Y_{s\wedge t}) - \int_0^{s\wedge t} Af(Y_u)du$$

Then if $r<s\leq t$ we have by (8.50) and (8.51)

$$(8.53)\qquad E^x[M_t J_s|\mathcal{N}_r] = E^x[M_s f(Y_s)|\mathcal{N}_r] - E^x[M_t\int_0^s Af(Y_u)du|\mathcal{N}_r]$$

$$= f(Y_r)E[M_t|\mathcal{N}_r] + E^x[\int_r^s M_u\cdot Af(Y_u)du|\mathcal{N}_r] - E^x[M_t\int_0^s Af(Y_u)du|\mathcal{N}_r]$$

$$= f(Y_r)E[M_t|\mathcal{N}_r] - E^x[M_t\int_0^r Af(Y_u)du|\mathcal{N}r]$$

$$= E^x[M_t J_r|\mathcal{N}_r] = J_r E^x[M_t|\mathcal{N}_r]$$

Now note that if $dS=M_t dR^x$, then the conditional expectations of a random variable Z wrt. a σ-algebra $\mathcal{H}$ and wrt. the two measures S and R^x are related by

$$(8.54)\qquad E^S[Z|\mathcal{H}]\cdot E^{R^x}[M_t|\mathcal{H}] = E^{R^x}[M_t Z|\mathcal{H}]$$

(To see this verify that, for all $H\in\mathcal{H}$,

$$\int_H E^S[Z|\mathcal{H}]M_t dR^x = \int_H Zds = \int_H ZM_t dR^x = \int_H E^{R^x}[M_t Z|\mathcal{H}]dR^x)$$

Applying (8.54) with $Z=J_s$, $\mathcal{H}=\mathcal{N}_r$ we get by (8.53) for $s>r$

$$E^S[J_s|\mathcal{N}_r]\cdot E^x[M_t|\mathcal{N}_r] = E^x[M_t J_s|\mathcal{N}_r] = J_r\cdot E^x[M_t|\mathcal{N}_r]$$

Since $E^x[M_t|\mathcal{N}_r] > 0$ a.s. we conclude that

$$E^S[J_s|\mathcal{N}_r] = J_r,$$

så J_s is a martingale for the measure S. By uniqueness of the martingale problem for A we conclude that $\tilde{S} = \tilde{Q}^x$, where $\tilde{S}$ is the measure on $(\mathbb{R}^n)^{[0,\infty)}$ corresponding to S under the identification $\omega(\cdot)\leftrightarrow Y^x_\cdot(\omega)$. In other words, we have established (8.48).

This result can be turned around to say that the law of $\{Y_s;s\leq t\}$ is absolutely continuous wrt. the law of $\{X_s;s\leq t\}$ if the drift term $a(t,\omega)$ has the form $a(t,\omega) = a(Y_t(\omega))$:

COROLLARY 8.23. Let X_t be as in Theorem 8.22 and let

(8.55) $dY_t = a(Y_t)dt + b(Y_t)dt + \sigma(Y_t)dB_t, \quad Y_0 = x$

where $a(y)$ satisfies the usual conditions (5.14), (5.15). Define

(8.56) $dW_t = -\frac{1}{2}(\sigma^{-1}(X_t)a(X_t))^2 dt + (\sigma^{-1}(X_t)a(X_t))^T dB_t, \quad W_0 = 0$

and

(8.57) $N_t = e^{W_t}$

Then

(8.58) $E^x[N_t f_1(X_{t_1})\cdots f_k(X_{t_k})] = E^x[f_1(Y_{t_1})\cdots f_k(Y_{t_k})]$

for all $0 \leq t_1 \leq \ldots \leq t_k \leq t, \quad f_i \in C_0^2$.

REMARK. A useful application of this result is that when X_t is an Ito diffusion with a diffusion coefficient matrix which has a bounded inverse, then the introduction of a drift will not change X_t dramatically. For example, X_t will have the same polar sets as before (see the definition preceding Theorem 9.12).

Consider the special case when $X_t = B_t$, n-dimensional Brownian motion. Then Theorem 8.22 says that the process

$$dY_t = d\hat{B}_t = a(t,\omega)dt + dB_t$$

becomes itself a Brownian motion with respect to the law $\hat{P}^x$ defined by

(8.59) $\hat{P}^x(F) = \int_F M_t dP^x \quad \text{for} \quad F \in \mathcal{N}_t,$

where

(8.60) $M_t = \exp\left(-\frac{1}{2}\int_0^t a^2(s,\omega)ds - \int_0^t a^T(s,\omega)dB_s\right)$

This transformation $P^x \to \hat{P}^x$ of measures is called the Cameron-Martin-Girsanov transformation. By proceeding as follows we see that it can be used to produce weak solutions of stochastic differential equations:

THEOREM 8.24. (The Cameron-Martin-Girsanov formula II)

Suppose

(8.61) $dX_t = b(X_t)dt + \sigma(X_t)dB_t, \; X_0 = x, \; b \in \mathbb{R}^n, \; \sigma \in \mathbb{R}^{n\times n}$

Define

(8.62) $$d\hat{B}_t = c(X_t)dt + dB_t,$$

where $c\colon R^n \to R^n$ is a bounded Borel function, and

(8.63) $$\hat{P}^x(F) = \int_F M_t dP^x \quad \text{for } F \in \mathcal{N}_t,$$

where $\mathcal{N}_t$ is the σ-algebra generated by $\{\hat{B}_s;\ s\leq t\}$ and

(8.64) $$M_t = \exp\left(-\tfrac{1}{2}\int_0^t c^2(X_s)ds - \int_0^t c^T(X_s)dB_s\right)$$

Then $\hat{B}_t$ is a Brownian motion wrt. $\hat{P}^x$ and (since $dB_t = d\hat{B}_t - c(X_t)dt$) we have

(8.65) $$dX_t = (b(X_t) - \sigma(X_t)c(X_t))dt + \sigma(X_t)d\hat{B}_t$$

Thus $(X_t, \hat{B}_t, \hat{P}^x)$ is a weak solution of (8.65) if X_t is a (weak or strong) solution of (8.61).

IX. Applications to Boundary Value Problems

A) The Dirichlet problem. We now use the preceding results to solve the following generalization of the Dirichlet problem stated in the introduction:

Given a domain $D \subset \mathbb{R}^n$, a semi-elliptic partial differential operator L on $C^2(W)$, where $W \supset \bar{D}$ is open, of the form

$$L = \sum a_{ij}(x) \frac{\partial^2}{\partial x_i \partial x_j} + \sum b_i(x) \frac{\partial}{\partial x_i}$$

and a continuous function ϕ on the boundary ∂D of D, find a function $\tilde{\phi}$ on D ("the L-harmonic extension of ϕ") such that

(9.1)

(i) $L\tilde{\phi} = 0$ in D

(ii) $\lim_{\substack{x \to y \\ x \in D}} \tilde{\phi}(x) = \phi(y)$ for all "reasonable" (i.e. regular) $y \in \partial D$

(L is called semi-elliptic (elliptic) when all the eigenvalues of the symmetric matrix $a_{ij}(x)$ are non-negative (positive), for all x). We will define regular points later. With this definition the Dirichlet problem turns out to have a (unique) solution under suitable conditions on L and D (See Theorems 9.11, 9.13 and 9.14).

The idea is the following: First we find an Ito diffusion $\{X_t\}$ whose characteristic operator $\mathcal{A}$ extends L. To obtain this we simply choose $\sigma \in \mathbb{R}^{n \times n}$ such that

(9.2) $$\tfrac{1}{2}\sigma\sigma^T = [a_{ij}]$$

We assume that σ and $b = [b_i]$ satisfy conditions (5.14), (5.15) of Theorem 5.5. (For example, if each $a_{ij} \in C^2$ is bounded and has bounded first and second partial derivatives, then such a square root σ can be found. See Fleming and Rishel [1, p. 123]). Next we let X_t be the solution of

(9.3) $$dX_t = b(X_t)dt + \sigma(X_t)dB_t,$$

where B_t is n-dimensional Brownian motion.

Then our candidate for the solution $\tilde{\phi}$ is

(9.4) $$\tilde{\phi}(x) = E^x[\phi(X_{\tau_D})].$$

The first problem we meet is that (9.4) only makes sense if

(9.5) $$Q^x[\tau_D < \infty] = 1 \quad \text{for all } x \in D.$$

From now on we assume that (9.5) holds.

For example, if L is elliptic in $\bar{D}$ and D is bounded, then this assumption is satisfied (Dynkin [3, p. 43]). We also see that it is satisfied for the process X_t associated to the parabolic L in Example 7.17. In Dynkin [3, p. 34-35], it is proved that (9.5) is a consequence of the weaker condition

(9.6) $\quad Q^x[\tau_D<\infty] > 0 \quad$ for all $x \in D$.

Next we define what we mean by regular points. First we will need the following useful results:

(As before we let $\mathcal{M}_t$ denote the σ-algebra generated by X_s; $s \leq t$).

LEMMA 9.1. (The 0-1 law) Let $H \in \bigcap_{t>0} \mathcal{M}_t$. Then either $Q^x(H) = 0$ or $Q^x(H) = 1$.

Proof. From the strong Markov property (7.15) we have

$$E^x[\theta_t\eta|\mathcal{M}_t] = E^{X_t}[\eta]$$

for all bounded, $\mathcal{M}$-measurable $\eta : \Omega \to \mathbb{R}$

This implies that

$$\int_H \theta_t\eta \cdot dQ^x = \int_H E^{X_t}[\eta]dQ^x, \quad \text{for all } t.$$

First assume that $\eta = \eta_k = g_1(X_{t_1})\cdots g_k(X_{t_k})$, where g_i is bounded and continuous. Then letting $t \to 0$ we obtain

$$\int_H \eta dQ^x = \lim_{t\to 0} \int_H \theta_t\eta dQ^x = \lim_{t\to 0} \int_H E^{X_t}[\eta]dQ^x = Q^x(H)E^x[\eta]$$

by Feller continuity (Lemma 8.4) and bounded convergence.

Approximating the general η by functions η_k as above we conclude that

$$\int_H \eta dQ^x = Q^x(H)E^x[\eta]$$

for all bounded $\mathcal{M}$-measurable η. If we put $\eta = \chi_H$ we obtain $Q^x(H) = (Q^x(H))^2$, which completes the proof.

COROLLARY 9.2. Let $y \in \mathbb{R}^n$. Then

$$\text{either} \quad Q^y[\tau_D=0] = 0 \quad \text{or} \quad Q^y[\tau_D=0] = 1.$$

Proof. $H = \{\omega; \tau_D=0\} \in \bigcap_{t>0} \mathcal{M}_t$.

In other words, either a.a. paths X_t starting from y stay within D for a positive period of time or a.a. paths X_t starting form y leave D immediately. In the last case we call the point y regular, i.e.

DEFINITION 8.3. A point $y \in \partial D$ is called regular for D (wrt. X_t) if

$$Q^y[\tau_D=0] = 1.$$

Otherwise the point y is called irregular.

EXAMPLE 9.4. Corollary 9.2 may seem hard to believe at first glance. For example, if X_t is a 2-dimensional Brownian motion B_t and D is the square $[0,1] \times [0,1]$ one might think that, starting from $(\frac{1}{2},0)$, say, half of the paths will stay in the upper half plane and half in the lower, for a positive period of time. However, Corollary 9.2 says that this is not the case: Either they all stay in D initially or they all leave D immediately. Symmetry considerations imply that the first alternative is impossible.

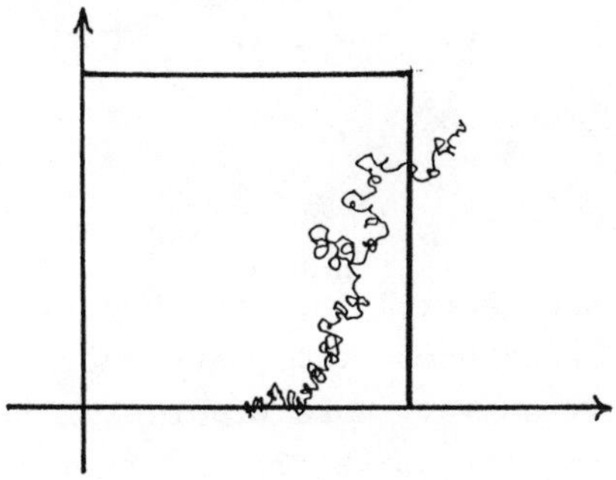

Thus $(\frac{1}{2},0)$, and similarly all the other points of ∂D, are regular for D wrt. B_t.

EXAMPLE 9.5. Let $D = [0,1] \times [0,1]$ and let L be the parabolic differential operator

$$Lf(t,x) = \frac{\partial f}{\partial f} + \frac{1}{2} \cdot \frac{\partial^2 f}{\partial x^2} \ ; \ (t,x) \in \mathbb{R}^2.$$

(See Example 7.17)

Here

$$b = \begin{bmatrix} 1 \\ 0 \end{bmatrix} \quad \text{and} \quad a = [a_{ij}] = \frac{1}{2} \begin{bmatrix} 0 & 0 \\ 0 & 1 \end{bmatrix}.$$

So, for example, if we choose $\sigma = \begin{bmatrix} 0 & 0 \\ 1 & 0 \end{bmatrix}$, we have $\frac{1}{2}\sigma\sigma^T = a$. This gives the following stochastic differential equation for the Ito diffusion X_t associated with L:

$$dX_t = \begin{bmatrix} 1 \\ 0 \end{bmatrix} dt + \begin{bmatrix} 0 & 0 \\ 1 & 0 \end{bmatrix} \begin{bmatrix} dB_t^{(1)} \\ dB_t^{(2)} \end{bmatrix}.$$

In other words,

$$X_t = \begin{bmatrix} t+t_0 \\ B_t \end{bmatrix}, \quad X_0 = \begin{bmatrix} t_0 \\ a \end{bmatrix}$$

where B_t is 1-dimensional Brownian motion.

So we end up with the graph of Brownian motion, which we started with in Example 7.17. In this case it is not hard to see that the irregular points of ∂D consist of the open line $\{0\} \times (0,1)$, the rest of the boundary points being regular.

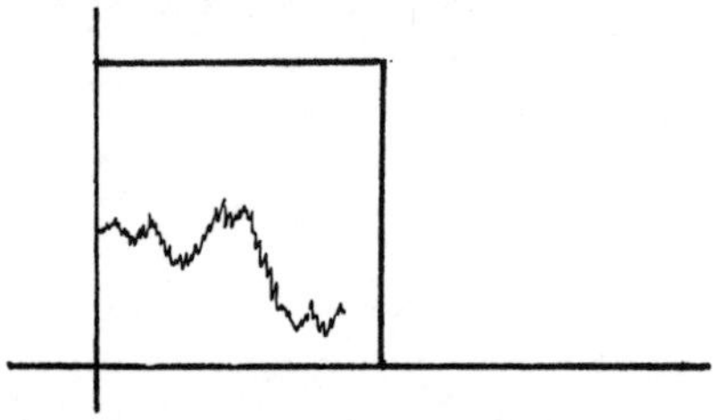

EXAMPLE 9.6. Let $\Delta = \{(x,y) ; x^2 + y^2 < 1\} \subset \mathbb{R}^2$ and let $\{\Delta_n\}$ be a sequence of disjoint open in Δ centered at $(2^{-n},0)$, respectively, $n = 1,2,\ldots$ Put

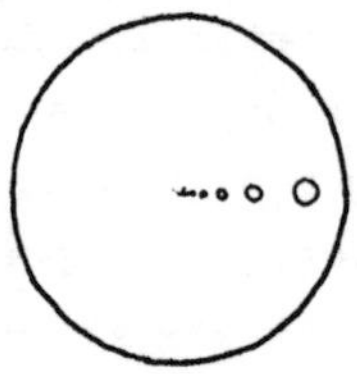

$$D = \Delta \setminus \bigcup_{n=1}^{\infty} \bar{\Delta}_n .$$

Then it is easy to see that all the points of $\partial\Delta \cup \bigcup_{n=1}^{\infty} \partial\Delta_n$ are regular for D wrt. 2-dimensional Brownian motion B_t, using a similar argument as in Example 9.4. But what about the point 0? The answer depends on the sizes of the discs Δ_n. More precisely, if r_n is the radius of Δ_n then 0 is a regular point for D if and only if

$$\sum_{n=1}^{\infty} \frac{n}{\log \frac{1}{r_n}} = \infty .$$

(This is a consequence of the famous Wiener criterion. See Port and Stone [1] p. 225)

Having defined regular points we now return to the Dirichlet problem and would like to conclude:

$$(9.1) \qquad \begin{aligned} &\text{(i)} \quad L\tilde{\phi} = 0 \quad \text{in } D \\ &\text{(ii)} \quad \lim_{\substack{x \to y \\ x \in D}} \tilde{\phi}(x) = \phi(y) \quad \text{for all regular } y \in \partial D. \end{aligned}$$

Unfortunately, this does not hold in general. Consider the following simple example:

EXAMPLE 9.7. Let $X(t) = (X_1(t), X_2(t))$ be the solution of the equations

$$dX_1(t) = dt$$
$$dX_2(t) = 0$$

so that $X(t) = X(0) + t(1,0) \in \mathbb{R}^2$; $t \geq 0$. Let

$$D = ((0,1) \times (0,1)) \cup ((0,2) \times (0,\tfrac{1}{2}))$$

and let ϕ be a continuous function on ∂D such that

$\phi = 1$ on $\{1\}\times[\tfrac{1}{2},1]$ and

$\phi = 0$ on $\{2\}\times[0,\tfrac{1}{2}]$.

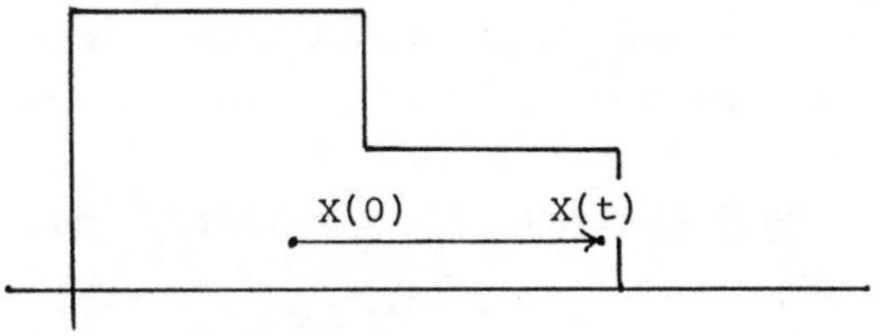

Then

$$\tilde{\phi}(t,x) = E^{t,x}[\phi(X_{\tau_D})] = \begin{cases} 1 & \text{if } x \in (\tfrac{1}{2},1) \\ 0 & \text{if } x \in (0,\tfrac{1}{2}), \end{cases}$$

so $\tilde{\phi}$ is not even continuous.

The point $y = (1,\tfrac{1}{2})$ is regular for X_t, but

$$\lim_{x\uparrow\frac{1}{2}} \tilde{\phi}(1,x) = 0 \neq \phi(1,\tfrac{1}{2}).$$

However, it is possible to formulate a weak, stochastic version of this problem. This stochastic version will always have a solution, which coincides with the (unique) solution of the original problem (9.1) is case such a solution exists. In the stochastic version we will replace (i) and (ii) in (9.1) by weaker requirements $(i)_s$ and $(ii)_s$. A natural candidate for the requirement $(i)_s$ would be that $\mathscr{A}\tilde{\phi} = 0$ in D. We will prove a slightly stronger property, which also makes it easy to obtain uniqueness: We will prove that $\tilde{\phi}$ is X-harmonic, in the sense that $\tilde{\phi}$ satisfies the mean value property wrt. X (see (7.19)):

DEFINITION 9.8. Let f be a locally bounded, measurable function on D. Then f is called X-harmonic in D if

$$f(x) = E^x[f(X_{\tau_U})]$$

for all $x \in D$ and all bounded open sets U with $\bar{U} \subset D$. We make two important observations:

LEMMA 9.9.

a) Let f be X-harmonic in D.
   Then $\mathscr{A}f = 0$ in D.

b) Conversely, suppose $f \in C^2(D)$ and $\mathscr{A}f = 0$ in D.
   Then f is X-harmonic

Proof. a) follows directly from the formula for $\mathscr{A}$.

b) follows from the Dynkin formula (Th. 7.10):

$$E^x[f(X_{\tau_U})] = \lim_{k\to\infty} E^x[f(X_{\tau_U\wedge k})] = f(x) + \lim_{k\to\infty} E^x[\int_0^{\tau_U\wedge k} (Lf)(X_s)ds] = f(x),$$

since $Lf = \mathscr{A}f = 0$ in U.

The most important examples of X-harmonic functions are given in the next result:

LEMMA 9.10. Let ϕ be a bounded measurable function on ∂D and put

$$\tilde{\phi}(x) = E^x[\phi(X_{\tau_D})] \; ; \; x \in D.$$

Then $\tilde{\phi}$ is X-harmonic. Thus, in particular, $\mathscr{A}\tilde{\phi} = 0$.

Proof. From (7.19) we have, if $\bar{V} \subset D$

$$\tilde{\phi}(x) = \int_{\partial V} \tilde{\phi}(y)\, Q^x[X_{\tau_V} \in dy] = E^x[\tilde{\phi}(X_{\tau_V})].$$

The stochastic Dirichlet problem:

Given a bounded measurable function ϕ on ∂D, find a function $\tilde{\phi}$ on D such that

$$\text{(9.7)} \quad \begin{array}{ll} (i)_s & \tilde{\phi} \text{ is X-harmonic} \\ (ii)_s & \lim_{t\uparrow\tau_D} \tilde{\phi}(X_t) = \phi(X_{\tau_D}) \quad \text{a.s } Q^x,\ x \in D. \end{array}$$

We first solve the stochastic Dirichlet problem (9.7) and then relate it to the original problem (9.1).

THEOREM 9.11. (Solution of the stochastic Dirichlet problem)

Let ϕ be a bounded measurable function on ∂D.

a) (Existence) Define

$$\text{(9.8)} \qquad \tilde{\phi}(x) = E^x[\phi(X_{\tau_D})]$$

Then $\tilde{\phi}$ solves the stochastic Dirichlet problem (9.7)

b) (Uniqueness) Suppose g is a bounded function on D such that

(1) g is X-harmonic

(2) $\lim_{t\uparrow\tau_D} g(X_t) = \phi(X_{\tau_D})$ a.s Q^x, $x \in D$.

Then $g(x) = E^x[\phi(X_{\tau_D})]$, $x \in D$.

Proof.

a) It follows from Lemma 9.10 that $(i)_s$ holds. Fix $x \in D$. Let $\{D_k\}$ be an increasing sequence of open sets such that $\bar{D}_k \subset D$ and $D = \bigcup_k D_k$. Put $\tau_k = \tau_{D_k}$, $\tau = \tau_D$. Then by the strong Markov property

$$\tilde{\phi}(X_{\tau_k}) = E^{X_{\tau_k}}[\phi(X_\tau)] = E^x[\theta_{\tau_k}(\phi(X_\tau))|\mathcal{F}_{\tau_k}]$$

$$(9.9) \qquad = E^x[\phi(X_\tau)|\mathcal{F}_{\tau_k}]$$

Now $M_k = E^x[\phi(X_\tau)|\mathcal{F}_{\tau_k}]$ is a bounded (discrete time) martingale so by the martingale convergence theorem (Appendix C) we get that

$$(9.10) \qquad \lim_{k\to\infty} \tilde{\phi}(X_{\tau_k}) = \lim_{k\to\infty} E^x[\phi(X_\tau)|\mathcal{F}_{\tau_k}] = \phi(X_\tau)$$

both pointwise for a.a.ω and in $L^p(Q^x)$, for all $p<\infty$. Moreover, by (9.9) it follows that for each k the process

$$N_t = \tilde{\phi}(X_{\tau_k \vee (t\wedge\tau_{k+1})}) - \tilde{\phi}(X_{\tau_k}) \; ; \; t \geq 0$$

is a martingale wrt. $\mathcal{G}_t = \mathcal{F}_{\tau_k \vee (t\wedge\tau_{k+1})}$.

So by the martingale inequality

$$(9.11) \qquad Q^x[\sup_{\tau_k \leq r \leq \tau_{k+1}} |\tilde{\phi}(X_r) - \tilde{\phi}(X_{\tau_k})| > \varepsilon] \leq \frac{1}{\varepsilon^2} E^x[|\tilde{\phi}(X_{\tau_{k+1}}) - \tilde{\phi}(X_{\tau_k})|^2]$$

$$\to 0 \text{ as } k\to\infty, \text{ for all } \varepsilon > 0.$$

From (9.10) and (9.11) we conclude that $(ii)_s$ holds.

b) Let D_k, τ_k be as in a). Then since g is X-harmonic we have

$$g(x) = E^x[g(X_{\tau_k})]$$

for all k. So by (2) and bounded convergence

$$g(x) = \lim_{k\to\infty} E^x[g(X_{\tau_k})] = E^x[\phi(X_{\tau_D})], \text{ as asserted.}$$

Finally we return to the original Dirichlet problem (9.1). First we establish that if a solution of this problem exists, it must be the solution of the stochastic Dirichlet problem found in Theorem 9.11, provided that X_t satisfies <u>Hunt's condition</u> (H):

(9.12) (H): Every semipolar set for X_t is polar for X_t.

A semipolar set is a countable union of thin sets and a measurable set $G \subset R^n$ is called thin (for X_t) if $Q^x[T_G=0] = 0$ for all x, where $T_G = \inf\{t>0; X_t \in G\}$ is the first hitting time of G. (Intuitively: For all starting points the process does not hit G immediately, a.s). A measurable set $F \subset R^n$ is called polar (for X_t) if $Q^x[T_G<\infty] = 0$ for all x. (Intuitively: For all starting points the process never hits F, a.s.). Clearly every polar set is semipolar, but the converse need not be true (consider the process in Example 9.7). However, condition (H) does hold for many important processes, for example Brownian motion (See Blumenthal and Getoor [1]).

We also need the following result, the proof of which can be found in Blumenthal and Getoor [1, Prop. II.3.3]:

LEMMA 9.12. Let $U \subset D$ be open and let I denote the set of irregular points of U. Then I is a semipolar set.

THEOREM 9.13. Suppose X_t satisfies Hunt's condition (H). Let ϕ be a bounded continuous function on ∂D. Suppose there exists a bounded $g \in C^2(D)$ such that

(i) $Lg = 0$ in D

(ii) $\lim_{\substack{x \to y \\ x \in D}} g(x) = \phi(y)$ for all regular $y \in \partial D$

Then $\quad g(x) = E^x[\phi(X_{\tau_D})]$.

Proof. Let $\{D_k\}$ be as in the proof Theorem 9.11. By Lemma 9.9b) g is X-harmonic and therefore

$$g(x) = E^x[g(X_{\tau_k})] \quad \text{for all } x \in D \text{ and all } k.$$

If $k \to \infty$ then $X_{\tau_k} \to X_{\tau_D}$ and so $g(X_{\tau_k}) \to \phi(X_{\tau_D})$ if X_{τ_D} is regular. From the Lemma 9.12 we know that the set I of irregular points of ∂D is semipolar. So by condition (H) the set I is polar and therefore $X_{\tau_D} \notin I$ a.s. Q^x. Hence

$$g(x) = \lim E^x[g(X_{\tau_k})] = E^x[\phi(X_{\tau_D})], \quad \text{as claimed.}$$

Under what conditions will the solution $\tilde{\phi}$ of the stochastic Dirichlet problem (9.7) also be a solution of the original Dirichlet problem (9.1)? This is a difficult question and we will content ourselves with the following partial answer:

THEOREM 9.14. Suppose L is uniformly elliptic in D, i.e. the eigenvalues of $[a_{ij}]$ are bounded away from 0 in D. (In addition we of course still assume that L is such that b and σ (given by (9.2)) satisfy (5.14) and (5.15). Let ϕ be a bounded continuous function on ∂D. Put

$$\tilde{\phi}(x) = E^x[\phi(X_{\tau_D})].$$

Then $\tilde{\phi}$ solves the Dirichlet problem (9.1), i.e.

(i) $L\tilde{\phi} = 0$ in D

(ii) $\lim_{\substack{x\to y\\ x\in D}} \tilde{\phi}(x) = \phi(y)$ for all regular $y\in\partial D$.

Proof. For k an integer and $\alpha>0$ we let $C^{k+\alpha}(G)$ denote the set of functions on G whose partial derivatives up to k'th order is Lipschitz continuous with exponent α. Choose an open ball Δ with $\bar{\Delta}\subset D$ and let $f\in C^\infty(\partial\Delta)$. Then, from the general theory of partial differential equations, for all $\alpha<1$ there exists a continuous function u on $\bar{\Delta}$ such that $u|\Delta\in C^{2+\alpha}(\Delta)$ and

(9.13) $$\begin{aligned} Lu &= 0 \quad \text{in } \Delta \\ u &= f \quad \text{on } \partial\Delta \end{aligned}$$

(see e.g. Dynkin [3, p. 226]. Since $u|\Delta\in C^{2+\alpha}(\Delta)$ we have: If K is any compact subset of Δ there exists a constant C only depending on K and the C^α-norms of the coefficients of L such that

(9.14) $$\|u\|_{C^{2+\alpha}(K)} \le C(\|Lu\|_{C^\alpha(\Delta)} + \|u\|_{C(\Delta)})$$

(See Bers, John and Schechter [1, Theorem 3 p. 232]) Combining (9.13) and (9.14) we obtain

(9.15) $$\|u\|_{C^{2+\alpha}(K)} \le C\|f\|_{C(\partial\Delta)}$$

By uniqueness (Theorem 9.13) we know that

(9.16) $$u(x) = \int f(y)\,d\mu_x(y),$$

where $d\mu_x = Q^x[X_{\tau_\Delta}\in dy]$ is the first exit distribution of X_t from Δ. From (9.15) it follows that

(9.17) $$\left|\int f d\mu_{x_1} - \int f d\mu_{x_2}\right| \le C\|f\|_{C(\partial\Delta)}|x_1-x_2|^\alpha \;;\; x_1,x_2\in K.$$

By approximating a given continuous function on $\partial\Delta$ uniformly by functions in $C^\infty(\partial\Delta)$ we see that (9.17) holds for all functions $f \in C(\partial\Delta)$. Therefore

(9.18) $$\|\mu_{x_1}-\mu_{x_2}\| \leq C|x_1-x_2|^\alpha \;;\; x_1,x_2 \in K$$

where $\| \ \|$ denotes the operator norm on measures on $\partial\Delta$. So if g is any bounded measurable function on $\partial\Delta$ we know that the function

$$\hat{g}(x) = \int g(y)\, d\mu_x(y) = E^x[g(X_{\tau_\Delta})]$$

belongs to the class $C^\alpha(K)$. Since $\tilde{\phi}(x) = E^x[\tilde{\phi}(X_{\tau_U})]$ for all open sets U with $\bar{U} \subset D$ and $x \in U$ (Lemma 9.10) this applies to $g = \tilde{\phi}$ and we conclude that $\tilde{\phi} \in C^\alpha(M)$ for any compact subset M of D.

We may therefore apply the solution to the problem (9.13) once more, this time with $f = \tilde{\phi}$ and this way we obtain that

$$u(x) = E^x[\tilde{\phi}(X_{\tau_D})] = \tilde{\phi}(x) \quad \text{belong to} \quad C^{2+\alpha}(M)$$

for any compact $M \subset D$. Therefore (i) holds by Lemma 9.9 b).

To obtain (ii) we apply a theorem from the theory of parabolic differential equations: The Kolmogorov backward equation

$$Lv = \frac{\partial v}{\partial t}$$

has a fundamental solution $v = p(t,x,y)$ jointly continuous in t,x,y for $t > 0$ and bounded in x,y for each fixed $t > 0$ (See Dynkin [3, Theorem 0.4 p. 227]). It follows (by bounded convergence) that the process X_t is a <u>strong Feller process</u>, in the sense that the function

$$x \to E^x[f(X_t)] = \int_{\mathbb{R}^n} f(y)p(t,x,y)\, dy$$

is continuous, for all $t > 0$ and all <u>bounded, measurable functions</u> f. In general we have:

(9.19) If X_t is a strong Feller Ito diffusion and $D \subset \mathbb{R}^n$ is open then

$$\lim_{\substack{x\to y\\ x\in D}} E^x[\phi(X_{\tau_D})] = \phi(y)$$

for all regular $y \in \partial D$ and bounded $\phi \in C(\partial D)$.

(See Theorem 13.3 p. 32-33 in Dynkin [3])

Therefore $\tilde{\phi}$ satisfies property (ii) and the proof is complete.

<u>REMARK</u>. One might ask why we don't require that

(ii)' $$\lim_{\substack{x\to y\\ x\in D}} \tilde{\phi}(x) = \phi(y) \quad \text{for } \underline{\text{all}} \quad y \in \partial D.$$

(and not just for regular points $y \in \partial D$) in our formulation (9.1) of the Dirichlet problem. The answer is that (ii)' is too much to hope for in general, even when L is elliptic: Consider Example 9.6 again, in the case when the point 0 is not regular. Choose $\phi \in C(\partial D)$ such that

$$\phi(0) = 1, \; 0 \leq \phi(y) < 1 \quad \text{for} \quad y \in \partial D \setminus \{0\}.$$

Since $\{0\}$ is polar for B_t (use the argument in Example 7.12) we have $B_{\tau_D} \neq 0$ a.s. and therefore

$$\tilde{\phi}(0) = E^0[\phi(B_{\tau_D})] < 1.$$

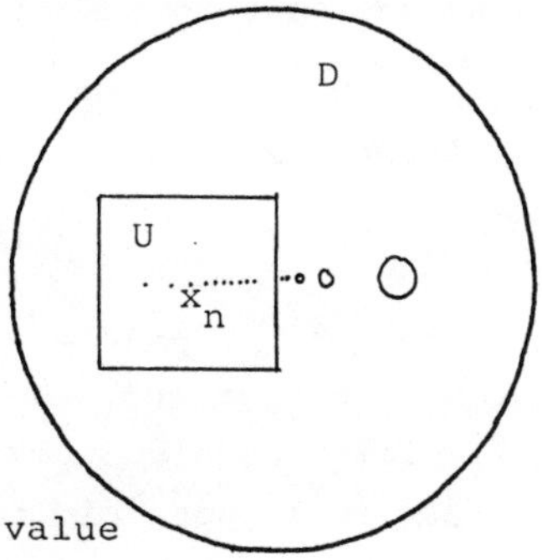

Let

$$U = \{(x,y) \in D;\ |x + \tfrac{1}{4}| < \tfrac{1}{4},\ |y| < \tfrac{1}{4}\}$$

and choose a sequence $x_n \in U \cap \mathbb{R}$ such that $x_n \to 0$. In this case we do not have $U \subset\subset D$ but (7.17) (and the mean value property (7.19)) still holds here, since the process X_t is Brownian motion.

Therefore, since 0 is a regular boundary point of U we have, as $n \to \infty$,

$$\tilde{\phi}(x_n) = E^{x_n}[\phi(B_{\tau_D})] = \int_{\partial U} E^y[\phi(B_{\tau_D})] \cdot Q^{x_n}[B_{\tau_U} \in dy]$$

$$\to \int_{\partial U} E^y[\phi(B_{\tau_D})]\, d\delta_0(y),$$

where δ_0 is the unit point mass at 0. So

$$\tilde{\phi}(x_n) \to \tilde{\phi}(0) = E^0[\phi(B_{\tau_D})] < 1 = \phi(0).$$

Therefore (ii)' does not hold in this case.

In general one can show that the regular points for Brownian motion are exactly the regular points in the classical potential theoretic sense, i.e. the points y on ∂D where the limit of the generalized Perron-Wiener-Brelot solution coincide with $\phi(y)$, for all $\phi \in C(\partial D)$. See Doob [1], Port and Stone [1] or Rao [1].

EXAMPLE 9.15. Let D denote the infinite strip

$$S = \{(t,x) \in \mathbb{R}^2 ; |x| < R\}, \quad \text{where } R > 0$$

and let L be the differential operator

$$Lf(t,x) = \frac{\partial f}{\partial t} + \frac{1}{2}\frac{\partial^2 f}{\partial x^2} ; \quad f \in C^2(D).$$

An Ito diffusion whose characteristic operator coincides with L is (see Example 8.5)

$$X_s = (s+t_0, B_s) ; \quad s \geq 0,$$

and all the points of ∂D are regular for this process. It is not hard to see that in this case (9.5) holds, i.e.

$$\tau_D < \infty \quad \text{a.s.}$$

So by Theorem 9.11 the function

$$\tilde{\phi}(t,x) = E^{t,x}[\phi(X_{\tau_D})]$$

is the solution of the stochastic Dirichlet problem (9.7), when $E^{t,x}$ denotes expection wrt. the probability law $Q^{t,x}$ for X starting at (t,x). Does $\tilde{\phi}$ also solve the problem (9.1)? Using the Laplace transform it is possible to find the distribution of the first exit point on ∂D for X, i.e. to find the distribution of the first time s that B_s reaches the value R. (See Karlin and Taylor [1, p. 363)]).

The result is

$$d\mu_R(s) = Q^{t,x}[\tau_D \in ds, B_{\tau_D} = R] = (R-x)(2\pi(s-t)^3)^{-\frac{1}{2}} \cdot \exp\left(\frac{-(R-x)^2}{2(s-t)}\right) ds,$$

and similarly for $B_{\tau_D} = -R$. Thus the solution $\tilde{\phi}$ may be written

$$\tilde{\phi}(t,x) = \int_{-\infty}^{\infty} \phi(s,R)\, d\mu_R(s) + \int_{-\infty}^{\infty} \phi(s,-R)\, d\mu_{-R}(s).$$

From the explicit formula for $d\mu_R$ and $d\mu_{-R}$ it is clear that $\tilde{\phi}$ is C^2 and we conclude that $L\tilde{\phi} = 0$ in D by Lemma 9.9. So $\tilde{\phi}$ satisfies (i) of (9.1). What about property (ii) of (9.1)? It is not hard to see that for $t > 0$

$$(9.20) \qquad E^{t_0,x}[f(X_t)] = (2\pi t)^{-\frac{1}{2}} \int_R f(t_0+t,y)\exp\left(-\frac{|x-y|^2}{2t}\right) dy$$

for all bounded, (t,x)-measurable functions f. (See (2.3)). Therefore X_t is <u>not</u> a strong Feller process, so we cannot appeal to (9.19) to obtain (ii). However, it is easy to verify directly that if

$|y| = R, t_1 > 0$ then for all $\varepsilon > 0$ there exists $\delta > 0$ such that $|x-y| < \delta$, $|t-t_1| < \delta \Rightarrow Q^{t,x}[\tau_D \in N] > 1-\varepsilon$, where $N = [t_1-\varepsilon, t_1+\varepsilon] \times \{y\}$. And this is easily seen to imply (ii).

REMARK. As the above example (and Example 9.7) shows, and Ito diffusion need not be a strong Feller process. However, we have seen that it is always a Feller process (Lemma 8.4).

B) The Poisson problem

Let $L = \sum a_{ij} \frac{\partial^2}{\partial x_i \partial x_j} + \sum b_i \frac{\partial}{\partial x_i}$ be a semi-elliptic partial differential operator on a domain $D \subset \mathbb{R}^n$ as before and let X_t be an associated Ito diffusion, described by (9.2) and (9.3). The (generalized) Poisson problem is the following:

Given a continuous function g on D find a C^2 function f in D such that

(9.21) $\quad Lf = -g$ in D

(9.22) $\quad \lim\limits_{\substack{x \to y \\ x \in D}} f(x) = 0$ for all regular $y \in \partial D$.

Again we will first study a stochastic version of the problem and then investigate the relation between the corresponding stochastic solution and the deterministic solution (if it exists) of (9.21-22):

THEOREM 9.16. (Solution of the stochastic Poisson problem)
Assume that

(9.23) $\quad E^x[\int_0^{\tau_D} |g(X_s)|ds] < \infty$ for all $x \in D$

(This occurs, for example, if g is bounded and $E^x[\tau_D] < \infty$ for all $x \in D$)
Define

(9.24) $\quad \check{g}(x) = E^x[\int_0^{\tau_D} g(X_s)ds]$

Then

(9.25) $\quad \mathcal{A}\check{g} = -g$ in D,

(9.26) $\quad \lim\limits_{t \uparrow \tau_D} \check{g}(X_t) = 0$ a.s. Q^x, for all $x \in D$.

Proof. Choose U open, $x \in U \subset\subset D$. Put $\eta = \int_0^{\tau_D} g(X_s)ds$, $\tau = \tau_U$.

Then by the strong Markov property (7.15)

$$\frac{E^x[\check{g}(X_\tau)]-\check{g}(x)}{E^x[\tau]} = \frac{1}{E^x[\tau]}(E^x[E^{X_\tau}[\eta]] - E^x[\eta])$$

$$= \frac{1}{E^x[\tau]}(E^x[E^x[\theta_\tau\eta|\mathscr{F}_\tau]] - E^x[\eta]) = \frac{1}{E^x[\tau]}(E^x[\theta_\tau\eta-\eta])$$

Approximate η by sums of the form

$$\eta^{(k)} = \sum g(X_{t_i})\chi_{\{t_i<\tau_D\}}\Delta t_i$$

Since

$$\theta_t\eta^{(k)} = \sum g(X_{t_i+t})\chi_{\{t_i+t<\tau_D^t\}}\Delta t_i \qquad \text{for all } k$$

(see the argument for (7.16)) we see that

$$\theta_\tau\eta = \int_\tau^{\tau_D} g(X_s)ds. \tag{9.27}$$

Therefore

$$\frac{E^x[\check{g}(X_\tau)]-\check{g}(x)}{E^x[\tau]} = \frac{-1}{E^x[\tau]}E^x[\int_0^\tau g(X_s)ds] \to -g(x) \quad \text{as} \quad U \downarrow x,$$

since g is continuous. This proves (9.25).

Put $H(x) = E^x[\int_0^{\tau_D} |g(X_s)|ds]$. Let D_k, τ_k be as in the proof of Theorem 9.11. Then by the same argument as above we get

$$E^x[H(X_{\tau_k\wedge t})] = E^x[\int_{\tau_k\wedge t}^{\tau_D} |g(X_s)|ds] \to 0 \quad \text{as} \quad t\to\tau_D, \quad k\to\infty$$

by dominated convergence. This implies (9.26).

REMARK. For functions g satisfying (9.23) define the operator $\mathscr{R}$ by

$$(\mathscr{R}g)(x) = \check{g}(x) = E^x[\int_0^{\tau_D} g(X_s)ds]$$

Then (9.25) can be written

$$(\mathscr{R}g) = -g \tag{9.28}$$

i.e. the operator $-\mathscr{R}$ is a right inverse of the operator $\mathscr{A}$. Similarly, if we define

$$\mathscr{R}_\alpha g(x) = E^x[\int_0^{\tau_D} g(X_s)e^{-\alpha s}ds] \quad \text{for} \quad \alpha > 0 \tag{9.29}$$

then the same method of proof as in Theorem 9.6 gives that

$$\mathcal{R}_\alpha g = -g + \alpha\mathcal{R}_\alpha g$$

or

(9.30) $$(\mathcal{A} - \alpha)\mathcal{R}_\alpha g = -g \quad ; \quad \alpha \geq 0$$

(If $\alpha > 0$ then the assumption on (9.23) can be replaced by the assumption that g is bounded (and continuous as before)).

Thus we may regard the operator $\mathcal{R}_\alpha$ as a generalization of the resolvent operator R_α discussed in Chapter VIII, and formula (9.30) as the analogue of Theorem 8.5 b).

Next we establish that if a solution f of original problem (9.21-22) exists, then f is the solution $\check{g}$ of the stochastic problem (9.25-26):

<u>THEOREM 9.17</u>. (Uniqueness theorem for the Poisson equation)
Assume that X_t satisfies Hunt's condition (H) ((9.12)). Assume that (9.23) holds and that there exists a function $f \in C^2(D)$ and a constant C such that

(9.31) $$|f(x)| \leq C(1+E^x[\int_0^{\tau_D} |g(X_s)|ds]) \quad \text{for all} \quad x \in D.$$

and

(9.32) $$Lf = -g \quad \text{in} \quad D,$$

(9.33) $$\lim_{x\to y} f(x) = 0 \quad \text{for all regular points} \quad y \in \partial D.$$

Then $f(x) = \check{g}(x) = E^x[\int_0^{\tau_D} g(X_s)ds]$.

<u>Proof</u>. Let D_k, τ_k be as in the proof of Theorem 9.11. Then by Dynkin's formula

$$E^x[f(X_{\tau_k})] - f(x) = E^x[\int_0^{\tau_k} (Lf)(X_s)ds] = -E^x[\int_0^{\tau_k} g(X_s)ds]$$

By dominated convergence we obtain

$$f(x) = \lim_{k\to\infty} (E^x[f(X_{\tau_k})]+E^x[\int_0^{\tau_k} g(X_s)ds]) = E^x[\int_0^{\tau_D} g(X_s)ds],$$

since X_{τ_D} is a regular point a.s. by condition (H) and Lemma 9.12.
Finally we combine the Dirichlet and Poisson problem and obtain the following result:

THEOREM 9.18. (Solution of combined stochastic Dirichlet and Poisson problem)

Let $\phi \in C(\partial D)$ be bounded and $g \in C(D)$ satisfy

(9.34) $$E^x[\int_0^{\tau_D} |g(X_s)|ds] < \infty \quad \text{for all } x \in D.$$

Define

(9.35) $$h(x) = E^x[\int_0^{\tau_D} g(X_s)ds] + E^x[\phi(X_{\tau_D})], \quad x \in D.$$

a) Then

(9.36) $$\mathcal{A}h = -g \quad \text{in } D$$

and

(9.37) $$\lim_{t\uparrow\tau_D} h(X_t) = \phi(X_{\tau_D}) \quad \text{a.s. } Q^x, \quad \text{for all } x \in D.$$

Moreover, if there exists a function $h_1 \in C^2(D)$ and a constant C such that

(9.38) $$|h_1(x)| < C(1+E^x[\int_0^{\tau_D} |g(X_s)|ds]), \quad x \in D,$$

and h_1 satisfies (9.36) and (9.37), then $h_1 = h$.

REMARK. With an approach similar to the one used in Theorem 9.13 one can prove that if L is uniformly elliptic in D and $g \in C^\alpha(D)$ (for some $\alpha > 0$) is bounded, then the function h given by (9.35) solves the Dirichlet-Poisson problem, i.e.

(9.39) $$Lh = -g \quad \text{in } D$$

and

(9.40) $$\lim_{\substack{x\to y\\ x\in D}} h(x) = \phi(y) \quad \text{for all regular } y \in \partial D.$$

The Green measure

The solution $\check{g}$ given by the formula (9.24) may be rewritten as follows:

DEFINITION 9.19. The Green measure (of X_t wrt. D at x), $G(x,\cdot)$ is defined by

(9.41) $$G(x,H) = E^x[\int_0^{\tau_D} \chi_H(X_s)ds], \quad H \subset \mathbb{R}^n \text{ Borel set}$$

or

$$(9.42) \qquad \int f(y)\, G(x,dy) = E^x\Big[\int_0^{\tau_D} f(X_s)ds\Big], \quad f \text{ bounded, continuous.}$$

In other words, $G(x,H)$ is the expected length of time the process stays in H before it exists from D. If X_t is Brownian motion, then

$$G(x,H) = \int_H G(x,y)dy,$$

where $G(x,y)$ is the classical Green function wrt. D and dy denotes Lebesque measure. (See Doob [1], Port and Stone [1] or Rao [1]).

Note that using the Fubini theorem we obtain the following relation between the Green measure G and the transition measure for X_t in D, $Q_t^D(x,H) = Q^x[X_t \in H,\ t<\tau_D]$:

$$(9.43) \qquad G(x,H) = E^x\Big[\int_0^\infty \chi_H(X_s)\cdot\chi_{[0,\tau_D)}(s)ds\Big] = \int_0^\infty Q_t^D(x,H)dt.$$

From (9.42) we get

$$(9.44) \qquad \check{g}(x) = E^x\Big[\int_0^{\tau_D} g(X_s)ds\Big] = \int_D g(y)G(x,dy),$$

which is the familiar formula for the solution of the Poisson equation in the classical case.

Also note that by using the Green function, we may regard the Dynkin formula as a generalization of the classical Green formula:

COROLLARY 9.20. (The Green formula)
Let $E^x[\tau_D] < \infty$ for all $x \in D$ and assume that $f \in C^2(D)\cap C(\bar{D})$ has bounded partial derivatives of order up to 2. Then

$$(9.45) \qquad f(x) = E^x[f(X_{\tau_D})] - \int_D (Lf)(y)G(x,dy)$$

In particular, if $f \in C_0^2(D)$ we have

$$(9.46) \qquad f(x) = -\int_D (Lf)(y)G(x,dy)$$

(As before $L = \sum b_i \frac{\partial}{\partial x_i} + \frac{1}{2}\sum(\sigma\sigma^T)_{ij}\frac{\partial^2}{\partial x_i \partial x_j}$ when

$dX_t = b(X_t)dt + \sigma(X_t)dB_t$)

Proof. By Dynkin's formula and (9.44) we have

$$E^x[f(X_{\tau_D})] = f(x) + E^x\left[\int_0^{\tau_D} (Lf)(X_s)ds\right] = f(x) + \int_D (Lf)(y)G(x,dy)$$

REMARK. Combining (9.28) with (9.46) we see that if $E^x[\tau_K] < \infty$ for all compacts $K \subset D$ and all $x \in D$, then $-\mathcal{R}$ is the inverse of the operator on $C_0^2(D)$:

(9.47) $\quad (\mathcal{R}f) = \mathcal{R}(\mathcal{A}f) = -f,$ for all $f \in C_0^2(D)$

More generally, for all $\alpha > 0$ we get the following analogue of Theorem 8.5:

(9.48) $\quad (\mathcal{A}-\alpha)(\mathcal{R}_\alpha f) = \mathcal{R}_\alpha(\mathcal{A}-\alpha)f = -f$ for all $f \in C_0^2(D)$

The first part of this is already established in (9.30) and the second part follows from the following useful extension of the Dynkin formula

(9.49) $\quad E^x[e^{-\alpha\tau} f(X_\tau)] = f(x) + E^x\left[\int_0^\tau e^{-\alpha s}(A-\alpha)f(X_s)ds\right].$

If $\alpha > 0$ this is valid for all stopping times $\tau \leq \infty$ and all $f \in C_0^2(\mathbb{R}^n)$. (To prove (9.49) use Ito's formula on $Y_t = e^{-\alpha t} f(Y_t)$ and proceed as in the proof of Dynkin's formula).

X. Application to Optimal Stopping

Problem 5 in the introduction is a special case of a problem of the following type:

PROBLEM 10.1. (The optimal stopping problem)
Let X_t be an Ito diffusion on $\mathbb{R}^n$ and let g (the reward function) be a given function on $\mathbb{R}^{n+1}$, satisfying

(10.1) a) $g(t,x) \geq 0$ for all $t \in \mathbb{R}$, $x \in \mathbb{R}^n$
b) g is continuous

Find a stopping time τ^* for $\{X_t\}$ such that

$$E^{(t,x)}[g(\tau^*,X_{\tau^*})] = \sup_\tau E^{(t,x)}[g(\tau,X_\tau)], \tag{10.2}$$

the sup being taken over all stopping times τ for $\{X_t\}$. We also want to find the corresponding optimal expected reward

$$g^*(t,x) = E^{(t,x)}[g(\tau^*,X_{\tau^*})]. \tag{10.3}$$

Here $g(\tau,X_\tau)$ is to be interpreted as 0 at the points $\omega \in \Omega$ where $\tau(\omega) = \infty$ and $E^{(t,x)}$ denotes the expectation operator with respect to the probability law $R^{(t,x)}$ of the process $Y_s=(t+s,X_s)$; $s \geq 0$ starting at $(t,x) \in \mathbb{R} \times \mathbb{R}^n$. This will be explained more precisely below.

We may regard X_s as the state of a game at time s, each ω corresponds to one sample of the game. For each time s we have the option of stopping the game, thereby obtaining the reward $g(s,X_s)$, or continue the game in the hope that stopping it at a later time will give a bigger reward. The problem is of course that we do not know what state the game is in at future times, only the probability distribution of the "future". Mathematically, this means that the possible "stopping" times we consider really are stopping times in the sense of Definition 7.3: The decision whether $\tau \leq s$ or not should only depend on the behaviour of the Brownian motion B_r (driving the process X) up to time s, or perhaps only on the behaviour of X_r up to time s. So, among all possible stopping times τ we are asking for the optimal one, τ^*, which gives the best result "in the long run", i.e. the biggest expected reward in the sense of (10.2).

In this chapter we will outline how a solution to this problem can be obtained using the material from the preceding chapter. Basically, the idea is the following:

STEP 1. For each $(t,x) \in \mathbb{R} \times \mathbb{R}^n$ and $s \geq 0$ let

$$Y_s^{(t,x)} = \begin{bmatrix} t+s \\ X_s^x \end{bmatrix} \tag{10.4}$$

i.e. $Y_s = Y_s^{(t,x)}$ is the graph of $X_s = X_s^x$, shifted to start at (t,x).

Then Y_s is an Ito diffusion.

STEP 2. Identify g^* with the least superharmonic majorant $\hat{g}$ of g wrt. Y_s.

STEP 3. Define the continuation region $D \subset \mathbb{R} \times \mathbb{R}^n$ by

$$D = \{(t,x) \; ; \; g(t,x) < g^*(t,x)\}.$$

Then (under certain conditions) the first exit time $\tau^* = \tau_D$ from D for Y_s solves the optimal stopping problem (10.2), and hence the optimal reward (10.3) is given by

$$g^*(t,x) = E^{(t,x)}[g(Y_{\tau_D})].$$

We now explain the individual steps:

STEP 1. Suppose $X_s = X_s^x$ solves the stochastic differential equation

$$dX_s = b(X_s)ds + \sigma(X_s)dB_s \; ; \; X_0 = x, \tag{10.5}$$

where $b \in \mathbb{R}^n$, $\sigma \in \mathbb{R}^{n \times m}$ and B_s is m-dimensional Brownian motion. Then $Y_s = Y_s^{(t,x)}$ given by (10.4) solves the equation

$$dY_s = \hat{b}(Y_s)ds + \hat{\sigma}(Y_s)dB_s, \quad Y_0 = \begin{bmatrix} t \\ x \end{bmatrix} \in \mathbb{R} \times \mathbb{R}^n \tag{10.6}$$

with

$$\hat{b} = \begin{bmatrix} 1 \\ \hline \bar{b} \end{bmatrix} \in \mathbb{R} \times \mathbb{R}^n, \quad \hat{\sigma} = \begin{bmatrix} 0 \; 0 \cdots 0 \\ \hline \bar{\sigma} \end{bmatrix} \in \mathbb{R}^{(n+1) \times m} \tag{10.7}$$

where $\bar{b}(z_0,\ldots,z_n) = b(z_1,\ldots,z_n)$, $\bar{\sigma}(z_0,\ldots,z_n) = \sigma(z_1,\ldots,z_n)$

So Y_s is an Ito diffusion.

The probability law of Y_s starting at (t,x) is the measure $R^{(t,x)}$ on Ω defined by

$$R^{(t,x)}(Y_{s_1} \in G_1,\ldots,Y_{s_k} \in G_k) = P^0(Y_{s_1}^{(t,x)} \in G_1,\ldots,Y_{s_k}^{(t,x)} \in G_k), \tag{10.8}$$

($G_1,\ldots,G_k$ Borel sets in $\mathbb{R} \times \mathbb{R}^n$)

where as before P^0 is the probability law of m-dimensional Brownian motion starting at 0. (See (7.7)). For each (t,x) this gives a measure $R^{(t,x)}$ on the σ-algebra $\mathcal{N}$ generated by $\{Y_s\}_{s\geq 0}$.

STEP 2. Step 1 really reduces the optimal stopping problem 10.1 to the case where the reward function does not depend on time (replacing $\{X_s\}$ by $\{Y_s\}$). So we consider the situation where we have an Ito diffusion $(\{Z_s\}_{s\geq 0}, \Omega, \mathcal{G}_t, \{P^z\}_{z\in R^k})$ with state space $\mathbb{R}^k$, on which a nonnegative continuous function $z \to g(z)$ is defined. The expectation operator wrt. P^z will be denoted by E^z. The optimal stopping problem then is to find an $\mathcal{F}_t$-stopping time τ^* such that

(10.9) $$E^z[g(Z_{\tau^*})] = \sup_{\tau} E^z[g(Z_\tau)],$$

the sup being taken over all $\mathcal{F}_t$-stopping times τ, and to find the corresponding optimal reward

(10.10) $$g^*(z) = E^z[g(Z_{\tau^*})]$$

As before we interpret $g(Z_\tau)$ as 0 whenever $\tau = \infty$.

A basic concept in the solution of (10.9), (10.10) is the following:

DEFINITION 10.2 A lower bounded measureable function $f: \mathbb{R}^k \to (-\infty,\infty]$ is called superharmonic (wrt. Z_s) if

(10.11) $$f(z) \geq E^z[f(Z_\tau)]$$

for all stopping times τ and all $z \in \mathbb{R}^k$, and in addition

(10.12) $$f(z) = \lim_{k\to\infty} E^z[f(Z_{\tau_k}]$$

for any sequence $\{\tau_k\}$ of stopping times such that $\tau_k \to 0$ a.s P^z, and for all $z \in \mathbb{R}^k$.

A (lower bounded measurable) function f which satisfies (10.11) is called supermeanvalued. Note that if f is super-meanvalued and

(10.13) $$f(z) \leq \underline{\lim}_{k\to\infty} E^z[f(Z_{\tau_k})],$$

then f is superharmonic, because (10.11) gives the other inequality. In particular, if f is supermeanvalued and lower semicontinuous, then f is superharmonic.

REMARK. If $f \in C^2(\mathbb{R}^k)$ it follows from Dynkin's formula that f is superharmonic wrt. Z_s if and only if

$$\hat{A}f \leq 0$$

where $\hat{A}$ is the generator of Z_s. This is often a useful criterion (See Example 10.15)

If $Z_s = B_s$ is the Brownian motion in $\mathbb{R}^k$ then the superharmonic functions for Z_s coincide with the (lower bounded) superharmonic functions in classical potential theory. See Doob [1] or Port and Stone [1].

We state some useful properties of superharmonic and supermeanvalued functions.

LEMMA 10.3

a) If f is superharmonic (supermeanvalued) and $\alpha > 0$, then αf is superharmonic (supermeanvalued).

b) If f_1, f_2 are superharmonic (supermeanvalued), then $f_1 + f_2$ is superharmonic (supermeanvalued)

c) If $\{f_j\}_{j\in J}$ is a family of supermeanvalued functions, then $f(z) = \inf_{j\in J}\{f_j(z)\}$ is supermeanvalued (J is any set).

d) If $f_1, f_2, \ldots$ are nonnegative superharmonic (supermeanvalued) functions and $f_k \uparrow f$ pointwise, then f is superharmonic (supermeanvalued)

e) If f is supermeanvalued and $\sigma \leq \tau$ are stopping times, then $E^z[f(Z_\sigma)] \geq E^z[f(Z_\tau)]$

f) If f is supermeanvalued and H is a Borel set, then $\tilde{f}(z) = E^z[f(Z_{\tau_H})]$ is supermeanvalued. If in addition f is lower semicontinuous, then $\tilde{f}$ is superharmonic.

Proof of Lemma 10.3

a) and b) are straightforward.

c) Suppose f_j is supermeanvalued for all $j \in J$. Then

$$f_j(z) \geq E^z[f_j(Z_\tau)] \geq E^z[f(Z_\tau)] \quad \text{for all } j,$$

so $f(z) = \inf f_j(z) \geq E^z[f(Z_\tau)]$, as required.

d) Suppose f_j is supermeanvalued, $f_j \uparrow f$. Then

$$f(z) \geq f_j(z) \geq E^z[f_j(Z_\tau)] \quad \text{for all } j, \text{ so}$$

$$f(z) \geq \lim_{j\to\infty} E^z[f_j(Z_\tau)] = E^z[f(Z_\tau)],$$

by monotone convergence. Hence f is supermeanvalued. If each f_j also satisfies (10.12) then

$$f_j(z) = \lim_{k\to\infty} E^z[f_j(Z_{\tau_k})] \leq \varliminf_{k\to\infty} E^z[f(Z_{\tau_k})]$$

and so

$$f(z) \leq \varliminf_{k\to\infty} E^z[f(Z_{\tau_k})].$$

e) If f is supermeanvalued we have by the Markov property when $s>t$

$$E^z[f(Z_s)\,|\,\mathcal{F}_t] = E^{Z_t}[f(Z_{s-t})] \leq f(Z_t), \tag{10.14}$$

i.e. the process

$$\zeta_t = f(Z_t)$$

is a supermartingale wrt. the σ-algebras $\mathcal{F}_t$ generated by $\{B_r; r\leq t\}$. (Appendix C). Therefore, by Doob's optional sampling theorem (see Gihman and Skorohod [3, Theorem 6 p.11]) we have

$$E^z[f(Z_\sigma)] \geq E^z[f(Z_\tau)]$$

for all stopping times σ, τ with $\sigma\leq\tau$ a.s. P^z.

f) Suppose f is supermeanvalued. By the strong Markov property (7.15) and formula (7.16) we have

$$E^z[\tilde{f}(Z_\tau)] = E^z[E^{Z_\tau}[f(Z_{\tau_H})]] = E^z[E^z[\theta_\tau f(Z_{\tau_H})\,|\,\mathcal{F}_\tau]]$$

(10.15)

$$= E^z[\theta_\tau f(Z_{\tau_H})] = E^z[f(Z_{\tau_H^\tau})]$$

where $\tau_H^\tau = \inf\{t>\tau;\ Z_t\notin H\}$. Since $\tau_H^\tau \geq \tau_H$ we have by e)

$$E^z[\tilde{f}(Z_\tau)] \leq E^z[f(Z_{\tau_H})] = \tilde{f}(z),$$

so $\tilde{f}$ is supermeanvalued.

From (10.15) and Fatou's lemma we see that if f is lower semicontinuous, then (10.13) holds for $\tilde{f}$:

$$\varliminf_{k\to\infty} E^z[\tilde{f}(Z_{\tau_k})] = \varliminf_{k\to\infty} E^z[f(Z_{\tau_H^{\tau_k}})]$$
$$\geq E^z[\varliminf_{k\to\infty} f(Z_{\tau_H^{\tau_k}})] \geq E^z[f(Z_{\tau_H})]$$
$$= \tilde{f}(z). \quad \text{So } \tilde{f} \text{ is superharmonic.}$$

The following concepts are fundamental:

DEFINITION 10.4. Let h be a real measurable function on $\mathbb{R}^k$. If f is a superharmonic (supermeanvalued) function and $f \geq h$ we say that f is a superharmonic (supermeanvalued) majorant of h (wrt. Z_s). The function

(10.16) $\hat{h}(z) = \inf_f f(z)$; $z \in \mathbb{R}^k$,

the inf being taken over all supermeanvalued majorants of h, is again supermeanvalued (Lemma 10.3 c)) and $\hat{h}$ is the least supermeanvalued majorant of h. If $\hat{h}$ is superharmonic then $\hat{h}$ is also the least superharmonic majorant of h.

Let $g \geq 0$ and let f be a supermeanvalued majorant of g. Then if τ is a stopping time

$$f(z) \geq E^z[f(z_\tau)] \geq E^z[g(Z_\tau)]$$

So

$$f(z) \geq \sup_\tau E^z[g(Z_\tau)] = g^*(z)$$

Therefore we always have

(10.17) $\hat{g}(z) \geq g^*(z)$ for all $z \in \mathbb{R}^k$

What is not so easy to see is that the converse inequality also holds, i.e. that in fact

$$\hat{g} = g^*$$

We will prove this after we have established a useful iterative procedure for calculating $\hat{g}$. Before we give such a procedure let us introduce a concept which is related to superharmonic functions:

DEFINITION 10.5. A measurable function $f : \mathbb{R}^k \to [0,\infty]$ is called excessive (wrt. Z_s) if

(10.18) $f(z) \geq E^z[f(Z_s)]$ for all $s \geq 0$, $z \in \mathbb{R}^k$.

and

(10.19) $f(z) = \lim_{s\to 0} E^z[f(Z_s)]$ for all $z \in \mathbb{R}^k$

It is clear that a nonnegative superharmonic function must be excessive. What is not so obvious, is that the converse also holds:

THEOREM 10.6. Let $f : \mathbb{R}^k \to [0,\infty]$. Then f is excessive wrt. Z_s if and only if f superharmonic wrt. Z_s.

Proof in a special case. Let L be the differential operator associated to Z as in (7.29), so that L coincides with the generator $\hat{A}$ of Z on C_0^2. We only prove the theorem in the special case when $f \in C^2(\mathbb{R}^k)$ and Lf is bounded: Then by Ito's formula /Dynkin's formula we have

$$E^z[f(Z_t)]=f(z)+E^z[\int_0^t Lf(Z_s)ds] \quad \text{for all} \quad t\geq 0,$$

so if f is excessive then $Lf\leq 0$. Therefore, if τ is a stopping time we get

$$E^z[f(Z_{t\wedge\tau})]\leq f(t) \quad \text{for all} \quad t\geq 0$$

Letting $t\to\infty$ we see that f is superharmonic.

A proof in the general case can be found in Dynkin [3, p.5].

The first iterative procedure for $\hat{g}$ is the following:

THEOREM 10.7. (Construction of the least superharmonic majorant). Let $g = g_0$ be a nonnegative, continuous function on $\mathbb{R}^k$ and define inductively

$$g_n(z) = \sup_{s\in S_n} E^z[g_{n-1}(Z_s)], \tag{10.20}$$

where $S_n= \{k \cdot 2^{-n};\ 0 \leq k \leq 4^n\}$, $n = 1, 2, \ldots$

Then $g_n\uparrow \hat{g}$ and $\hat{g}$ is the least superharmonic majorant of g.

Proof. Define $\check{g}(z) = \lim\limits_{n\to\infty} g_n(z)$.

Then

$$\check{g}(z) \geq g_n(z) \geq E^z[g_{n-1}(Z_s)] \quad \text{for all } n \text{ and all } s \in S_n.$$

Hence

$$\check{g}(z) \geq \lim_{n\to\infty} E^z[g_{n-1}(Z_s)] = E^z[\check{g}(Z_s)] \tag{10.21}$$

$$\text{for all} \quad s \in S = \bigcup_{n=1}^{\infty} S_n.$$

Since $\check{g}$ is a non-decreasing limit of lower semi-continuous functions (Lemma 8.4) $\check{g}$ is lower semicontinuous. Fix $s \in R$ and choose $s_k \in S$ such that $s_k \to s$. Then by (10.21), the Fatou lemma and lower semicontinuity

$$\check{g}(z) \geq \varliminf_{k\to\infty} E^z[\check{g}(Z_{s_k})] \geq E^z[\varliminf_{k\to\infty} \check{g}(Z_{s_k})] \geq E^z[\check{g}(Z_s)].$$

So $\check{g}$ is an excessive function. Therefore $\check{g}$ is superharmonic by Theorem 10.6 and hence $\check{g}$ is a superharmonic majorant of g. On the other hand, if f is any supermeanvalued majorant of g, then clearly by induction

$$f(z) \geq g_n(z) \quad \text{for all } n$$

and so $f(z) \geq \check{g}(z)$.

This proves that $\check{g}$ is the least supermeanvalued majorant $\hat{g}$ of g, and $\check{g}$ is superharmonic.

It is a consequence of Theorem 10.7 that we may replace the finite sets S_n by the whole interval $[0,\infty]$:

COROLLARY 10.8. Define $h_0 = g$ and inductively

$$h_n(z) = \sup_{s\geq 0} E^z[h_{n-1}(Z_s)] \; ; \; n = 1,2,\ldots$$

Then $h_n \uparrow \hat{g}$.

Proof. Let $\bar{h} = \lim h_n$. Then clearly $\bar{h} \geq \check{g} = \hat{g}$. On the other hand, since $\hat{g}$ is excessive we have

$$\hat{g}(z) \geq \sup_{s\geq 0} E^z[\hat{g}(Z_s)]$$

so by induction

$$\hat{g} \geq h_n \quad \text{for all } n.$$

Thus $\hat{g} = \bar{h}$ and the proof is complete.

STEP 3 AND COMPLETION OF STEP 2

We now apply these results to the diffusion $Z_s = Y_s = (t+s, X_s)$ with probability law $S^z = R^{(t,x)}$ when $z=y=(t,x)$ and obtain the following solution to the optimal stopping problem, basically due to Dynkin [1] (and, in a martingle context, Snell [1]):

THEOREM 10.9. (Existence theorem for optimal stopping)
Let g^* denote the optimal reward and $\hat{g}$ the least superharmonic majorant of the continuous reward function g.

a) Then

(10.22) $$g^*(t,x) = \hat{g}(t,x)$$

b) For $\varepsilon > 0$ let

$$D_\varepsilon = \{(t,x) \; ; \; g(t,x) < \hat{g}(t,x) - \varepsilon\}.$$

Suppose g is bounded. Then stopping at the first time τ_ε of exit from D_ε is close to being optimal, in the sense that

(10.23) $$|g^*(t,x) - E^{(t,x)}[g(Y_{\tau_\varepsilon})]| < 2\varepsilon$$

for all (t,x).

c) For arbitrary continous g let

$D = \{(t,x) \; ; \; g(t,x) < g^*(t,x)\}$, the continuation region.
For $N = 1,2,\ldots$ define $g_N = g \wedge N$, $D_N = \{(t,x); \; g_N(t,x) < (\hat{g}_N)(t,x)\}$ and $\sigma_N = \tau_{D_N}$. Then $D_N \subset D_{N+1}$, $D_N \subset D \cap g^{-1}([0,N))$, $D = \bigcup_N D_N$.
If $\sigma_N < \infty$ a.s. $R^{(t,x)}$ for all N then

(10.24) $$g^*(t,x) = \lim_{N\to\infty} E^{(t,x)}[g(Y_{\sigma_N})].$$

In particular, if $\tau_D < \infty$ a.s. $R^{(t,x)}$ and the family

$$\{g(Y_{\sigma_N})\}_N$$

is uniformly integrable wrt. $R^{(t,x)}$ (Appendix C), then

$$g^*(t,x) = E^{(t,x)}[g(Y_{\tau_D})]$$

and $\tau^* = \tau_D$ is an optimal stopping time.

Proof. First assume that g is bounded and define

$$\tilde{g}_\varepsilon(y) = E^y[\hat{g}(Y_{\tau_\varepsilon})] \quad \text{where } y=(t,x).$$

Then $\tilde{g}_\varepsilon$ is superharmonic by Lemma 10.3 f). We claim that

(10.25) $$g(y) \leq \tilde{g}_\varepsilon(y) + \varepsilon \quad \text{for all } y.$$

To see this suppose

$$(10.26)\quad \beta = \sup_y \{g(y) - \tilde{g}_\varepsilon(y)\} > \varepsilon.$$

Then for all $\eta > 0$ we can find y_0 such that

$$(10.27)\quad g(y_0) - \tilde{g}_\varepsilon(y_0) \geq \beta - \eta.$$

On the other hand, since $\tilde{g}_\varepsilon + \beta$ is a supermeanvalued majorant of g, we have

$$(10.28)\quad \hat{g}(y_0) \leq \tilde{g}_\varepsilon(y_0) + \beta.$$

Combining (10.27) and (10.28) we get

$$(10.29)\quad \hat{g}(y_0) \leq g(y_0) + \eta$$

Consider the two possible cases:

Case 1: $\tau_\varepsilon > 0$ a.s. R^{y_0}. Then by (10.29) and the definition of D_ε

$$g(y_0) + \eta \geq \hat{g}(y_0) \geq$$

$$E^{y_0}[\hat{g}(Y_{t\wedge\tau_\varepsilon})] \geq E^{y_0}[(g(Y_t) + \varepsilon)\chi_{\{t<\tau_\varepsilon\}}]$$

$$\to g(y_0) + \varepsilon \quad \text{as} \quad t \downarrow 0$$

This is a contradiction if $\eta < \varepsilon$.

Case 2: $\tau_\varepsilon = 0$ a.s R^{y_0}. Then $\tilde{g}_\varepsilon(y_0) = \hat{g}(y_0)$, so $g(y_0) \leq \tilde{g}_\varepsilon(y_0)$, contradicting (10.27) for $\eta < \beta$.

Therefore (10.26) leads to a contradiction.

Thus (10.25) is proved and we conclude that $\tilde{g} + \varepsilon$ is a superharmonic majorant of g. Therefore

$$(10.30)\quad \hat{g} \leq \tilde{g}_\varepsilon + \varepsilon = E[\hat{g}(Y_{\tau_\varepsilon})] + \varepsilon \leq E[(g+\varepsilon)(Y_{\tau_\varepsilon})] + \varepsilon \leq g^* + 2\varepsilon$$

and since ε was arbitrary we have by (10.17)

$$\hat{g} = g^*$$

If g is not bounded, let

$$g_N = \min(N,g), \quad N = 1,2,\ldots$$

Then

$$g^* \geq (g_N)^* = (g_N)^\wedge \uparrow h \quad \text{as} \quad n \to \infty, \quad \text{where} \quad h \geq \hat{g}$$

since h is a superharmonic majorant of g. Thus $h = \hat{g} = g^*$ and this proves (10.22) for general g. From (10.30) and (10.22) we obtain (10.23).

Finally, to obtain c) let us again first assume that g is bounded. Then, since

$$\tau_\varepsilon \uparrow \tau_D \quad \text{as} \quad \varepsilon \downarrow 0$$

and $\tau_D < \infty$ a.s we have

$$E^y[g(Y_{\tau_\varepsilon})] \to E^y[g(Y_{\tau_D})] \quad \text{as} \quad \varepsilon \downarrow 0,$$

and hence by (10.30) and (10.22)

(10.31) $\quad g^*(y) = E^y[g(Y_{\tau_D})]$ if g is bounded.

Finally, if g is not bounded define

$$h = \lim_{N\to\infty} (g_N)^\wedge .$$

Then h is superharmonic by Lemma 10.3 d) and since $(g_N)^\wedge \leq \hat{g}$ for all N we have $h \leq \hat{g}$. On the other hand $g_N \leq (\hat{g}_N) \leq h$ for all N and therefore $g \leq h$. Since $\hat{g}$ is the least superharmonic majorant of g we conclude that

(10.32) $\quad h = \hat{g}$

Hence by (10.31-32) we obtain (10.24):

$$g^*(y) = \lim_{N\to\infty}(g_N)^\wedge(y) = \lim_{N\to\infty} E^y[g_N(Y_{\sigma_N})] \leq \lim_{N\to\infty} E^y[g(Y_{\sigma_N})] \leq g^*(y).$$

Note that $(g_N)^\wedge \leq N$ everywhere, so if $g_N(y) < (g_N)^\wedge(y)$ then $g_N(y) < N$ and therefore $g(y) = g_N(y) < (g_N)^\wedge(y) \leq \hat{g}(y)$ and $g_{N+1}(y) = g_N(y) < (g_N)^\wedge(y) \leq (g_{N+1})^\wedge(y)$. Hence $D_N \subset D \cap \{y; g(y) < N\}$ and $D_N \subset D_{N+1}$ for all N. So by (10.32) we conclude that D is the increasing union of the sets D_N ; $N = 1,2,\ldots$ Therefore

$$\tau_D = \lim_{N\to\infty} \sigma_N .$$

So by (10.24) and uniform integrability we have

$$\hat{g}(y) = \lim_{N\to\infty}(g_N)^\wedge(y) = \lim_{N\to\infty} E^y[g_N(Y_{\sigma_N})] = E^y[\lim_{N\to\infty} g_N(Y_{\sigma_N})] = E^y[g(Y_{\tau_D})],$$

and the proof of Theorem 10.9 is complete.

The following consequence of Theorem 10.9 is often useful:

COROLLARY 10.10. Suppose there exists a Borel set H such that

$$\tilde{g}_H(t,x) = E^{(t,x)}[g(Y_{\tau_H})]$$

is a supermeanvalued majorant of g. Then

$$g^*(t,x) = \tilde{g}_H(t,x), \quad \text{so} \quad \tau^* = \tau_H \text{ is optimal.}$$

Proof. If $\tilde{g}_H$ is a supermeanvalued majorant of g then clearly

$$\hat{g}(t,x) \leq \tilde{g}_H(t,x)$$

On the other hand we of course have

$$\tilde{g}_H(t,x) \leq \sup_\tau E^{(t,x)}[g(Y_\tau)] = g^*(t,x),$$

so $g^* = \tilde{g}_H$ by Theorem 10.9 a).

COROLLARY 10.11. Let

$$D = \{(t,x);\ g(t,x) < \hat{g}(t,x)\}$$

and put

$$\tilde{g}(t,x) = \tilde{g}_D(t,x) = E^{(t,x)}[g(Y_{\tau_D})].$$

If $\tilde{g} \geq g$ then $\tilde{g} = g^*$.

Proof. Since $Y_{\tau_D} \notin D$ we have $g(Y_{\tau_D}) \geq \hat{g}(Y_{\tau_D})$ and therefore $g(Y_{\tau_D}) = \hat{g}(Y_{\tau_D})$, a.s. $R^{t,x}$. So $\tilde{g}(t,x) = E^{(t,x)}[\hat{g}(Y_{\tau_D})]$ is supermeanvalued since $\hat{g}$ is, and the result follows from Corollary 10.10.

Theorem 10.9 gives a sufficient condition for the existence of an optimal stopping time τ^*. Unfortunately, τ^* need not exist in general. For example, if

$$X_t = t \quad \text{for} \quad t \geq 0 \quad \text{(deterministic)}$$

and

$$g(t,x) = \frac{x^2}{1+x^2}\ ;\ x \in \mathbb{R}$$

then $g^*(t,x) = 1$, but there is no stopping time τ such that

$$E^{t,x}[g(X_\tau)] = 1.$$

However, we can prove that if an optimal stopping time τ^* exists, then it is of the form given in Theorem 10.9:

THEOREM 10.12. (Uniqueness theorem for optimal stopping)
Define as before

$$D = \{(t,x) \; ; \; g(t,x) < g^*(t,x)\} \subset \mathbb{R} \times \mathbb{R}^n .$$

Suppose there exists an optimal stopping time τ^* for the problem (10.1). Then

(10.33) $\quad \tau^* \geq \tau_D,$

(where τ_D is the first time from D of Y_s), and

(10.34) $\quad g^*(t,x) = E^{(t,x)}[g(Y_{\tau_D})] \quad$ for all (t,x).

Hence τ_D is an optimal stopping time for problem (10.1)

Proof. Let τ be a stopping time for Y_s and assume $R^{(t,x)}[\tau<\tau_D] > 0$. Since $g(Y_\tau) < g^*(Y_\tau)$ if $\tau < \tau_D$ and $g \leq g^*$ always, we have

$$E^{(t,x)}[g(Y_\tau)] = \int_{t<\tau_D} g(Y_\tau)dR^{(t,x)} + \int_{\tau\geq\tau_D} g(Y_\tau)dR^{(t,x)}$$

$$< \int_{\tau<\tau_D} g^*(Y_\tau)dR^{(t,x)} + \int_{\tau\geq\tau_D} g^*(Y_\tau)dR^{(t,x)}$$

$$= E^{(t,x)}[g^*(Y_\tau)] \leq g^*(t,x)$$

since g^* is superharmonic. This proves (10.33). To obtain (10.34) we note that since $\hat{g}$ is superharmonic we have by Lemma 10.3 e)

$$g^* = E[g(Y_{\tau^*})] \leq E[\hat{g}(Y_{\tau^*})] \leq E[\hat{g}(Y_{\tau_D})]$$

$$= E[g(Y_{\tau_D})] \leq g^* .$$

REMARK. The following observation is sometimes useful: Let $\mathscr{A}$ be the characteristic operator of X, so that

(10.35) $\quad (\hat{\mathscr{A}}f)(t,x) = \frac{\partial f}{\partial t} + \mathscr{A} f_t, \quad$ where $f_t(x) = f(t,x)$

is the characteristic operator of Y.
Let $g \in C^2(\mathbb{R}^{n+1})$. Then

(10.36) $\quad U = \{(t,x); \hat{\mathscr{A}}g(t,x)>0\} \subset \{(t,x); g(t,x)<g^*(t,x)\} = D.$

Consequently, from (10.33) we conclude that it is <u>never optimal to stop the process before it exists from</u> U. (But there may be cases when $U \neq D$, so that it is optimal to proceed beyond U before stopping. See Example 10.15).

To prove (10.36) choose $(t,x) \in U$ and let τ_0 be the first exit time from a bounded open set $W \ni (t,x)$, $W \subset U$. Then by Dynkin's formula, for $u > 0$

$$E^{(t,x)}[g(Y_{\tau_0 \wedge u}] = g(t,x) + E^{t,x}[\int_0^{\tau_0 \wedge u} \hat{A}\, g(Y_s)ds] > g(t,x)$$

so $g(t,x) < g^*(t,x)$.

EXAMPLE 9.13. Let $X_t = B_t$ be a Brownian motion in $\mathbb{R}^2$ and suppose that the reward function $g(t,x)$ does not depend on t:

$$g(t,x) = g(x) \geq 0.$$

Since g does not depend on t, the least superharmonic majorant of g wrt. Y_s coincides with the least superharmonic majorant of g wrt. B_t, i.e. the least superharmonic majorant in the classical sense. Using that B_t is recurrent in $\mathbb{R}^2$ (Example 7.11) it is not hard to show that the only nonnegative superharmonic functions in $\mathbb{R}^2$ are the constants.

Therefore

$$g^*(x) = \|g\|_\infty = \sup\{g(y)\ ;\ y \in \mathbb{R}^2\} \quad \text{for all } x.$$

So if g is unbounded $g^* = \infty$ and no optimal stopping exists. Assume therefore that g is bounded. The continuation region is

$$D = \{(t,x)\ ;\ g(x) < \|g\|_\infty\},$$

so if D is a polar set (i.e. cap $(\partial D) = 0$, where cap denotes the logarithmic capacity), then $\tau_D = \infty$ a.s. so no optimal stopping exists. (See Port and Stone [1]). On the other hand, if cap$(\partial D) > 0$ then $\tau_D < \infty$ a.s. and

$$E^x[g(B_{\tau_D})] = \|g\|_\infty = g^*,$$

so $\tau^* = \tau_D$ is optimal.

EXAMPLE 10.14. The situation is different in $\mathbb{R}^n$ for $n \geq 3$.

a) To illustrate this let $X_t = B_t$ be Brownian motion in $\mathbb{R}^3$ and let the reward function be

$$g(t,x) = g(x) = \begin{cases} |x|^{-1} & \text{for } |x| \geq 1 \\ 1 & \text{for } |x| < 1 \end{cases}$$

Then g is superharmonic (in the classical sense) in $\mathbb{R}^3$, so $g^* = g$ everywhere and the best policy is to stop immediately, no matter where the starting point is.

b) Let us change g to

$$h(t,x) = h(x) = \begin{cases} |x|^{-\alpha} & \text{for } |x| \geq 1 \\ 1 & \text{for } |x| < 1 \end{cases}$$

for some $\alpha > 1$. Let $H = \{x; |x|>1\}$ and define

$$\tilde{h}(x) = E^x[h(B_{\tau_H})] = P^x[\tau_H < \infty]$$

Then by Example 7.11

$$\tilde{h}(x) = \begin{cases} 1 & \text{if } |x| \leq 1 \\ |x|^{-1} & \text{if } |x| > 1, \end{cases}$$

i.e. $\tilde{h} = g$, which is superharmonic majorant of h. Therefore by Corollary 10.10

$$h^* = \tilde{h} = g,$$

$H = D$ and $\tau^* = \tau_H$ is an optimal stopping time.

<u>EXAMPLE 10.15</u>. Let $X_t = B_t$ be 1-dimensional Brownian motion and let the reward function be

$$g(t,x) = e^{-\alpha t+\beta x}$$

where $\alpha, \beta \geq 0$ are constants.
The characteristic operator $\mathscr{A}$ of $Y_s^{t,x} = \begin{bmatrix} t+s \\ B^x \end{bmatrix}$ is given by

$$\mathscr{A}f = \frac{\partial f}{\partial t} + \frac{1}{2}\cdot\frac{\partial^2 f}{\partial x^2}\ ; \quad f \in C^2.$$

(See Example 7.17).
Thus

$$\mathscr{A}g = (-\alpha + \tfrac{1}{2}\beta^2)g,$$

so if $\beta^2 \leq 2\alpha$ then $g^* = g$ and the best policy is to stop immediately. If $\beta^2 > 2\alpha$ we can use Theorem 10.7 to prove that $g^* = \infty$:

$$\sup_{s\in S_n} E^{(t,x)}[g(Y_s)] = \sup_{s\in S_n} E[e^{-\alpha(t+s) + \beta B_s^x}]$$

$$= \sup_{s\in S_n} [e^{-\alpha(t+s)} \cdot e^{\beta x + \frac{1}{2}\beta^2 s}] \qquad \text{(see the remark following (5.6))}$$

$$= \sup_{s\in S_n} g(t,x)\cdot e^{(-\alpha+\frac{1}{2}\beta^2)s} = g(t,x)\cdot \exp((-\alpha + \tfrac{1}{2}\beta^2)2^n),$$

so $g_n(t,x) \to \infty$ as $n \to \infty$.

Hence no optimal stopping exists in this case.

EXAMPLE 10.16. (When is the right time sell the stocks?)
We now return to a specified version of Problem 5 in the introduction:
Suppose the price X_t at time t of a person's assets (e.g. a house, stocks, oil...) varies according to a stochastic differential equation of the form

$$dX_t = rX_t dt + \alpha X_t dB_t, \quad X_0 = x > 0,$$

where B_t is 1-dimensional Brownian motion and r, α are known constants. (The problem of estimating r and α from a series of observations can be approached using filtering theory. See Chapter VI). Suppose that connected to the sale of the assets there is a fixed fee/tax or transaction cost $a > 0$. Then if the person decides to sell at time t the discounted net of the sale is

$$e^{-\rho t}(X_t - a),$$

where $\rho > 0$ is a given discounting factor. The problem is to find a stopping time τ that maximizes

$$E^{(t,x)}[e^{-\rho\tau}(X_\tau - a)] = E^{(t,x)}[g(\tau, X_\tau)],$$

where

$$g(t,x) = e^{-\rho t}(x-a).$$

The characteristic operator $\mathscr{A}$ of the process $Y_s = (t+s, X_s)$ is given by

$$\mathscr{A}f(t,x) = \frac{\partial f}{\partial t} + rx\,\frac{\partial f}{\partial x} + \frac{1}{2}\alpha^2 x^2 \frac{\partial^2 f}{\partial x^2}\,; \quad f \in C_0^2(\mathbb{R}^2)$$

Hence $\mathscr{A}g = -\rho e^{-\rho t}(x-a) + rxe^{-\rho t} = e^{-\rho t}((r-\rho)x+\rho a)$. So

$$U = \{(t,x);\ \mathscr{A}g(t,x) > 0\} = \begin{cases} \mathbb{R} \times \mathbb{R}_+ & \text{if } r \geq \rho \\ \{(t,x);\ x < \frac{a\rho}{\rho - r}\} & \text{if } r < \rho \end{cases}$$

So if $r \geq \rho$ we have $U = D = \mathbb{R} \times \mathbb{R}_+$ so τ^* does not exist. If $r > \rho$ then $g^* = \infty$ while if $r = \rho$ then

$$g^*(t,x) = xe^{-\rho t}$$

(the proofs of these statements are left as an exercise)

It remains to examine the case $r < \rho$.

First we establish that the region D must be invariant wrt. t, in the sense that

(10.37) $D + (t_0,0) = D$ for all t_0.

To prove (10.37) consider

$$D + (t_0,0) = \{(t+t_0,x);\ (t,x) \in D\} = \{(s,x);\ (s-t_0,x) \in D\}$$

$$=\{(s,x); g(s-t_0,x)<g^*(s-t_0,x)\}=\{(s,x); e^{\rho t_0}g(s,x)< e^{\rho t_0}g^*(s,x)\}$$

$$= \{(s,x); g(s,x)<g^*(s,x)\} = D,$$

where we have used that

$$g^*(s-t_0,x) = \sup_\tau E^{(s-t_0,x)}[e^{-\rho\tau}(X_\tau-a)] = \sup_\tau E[e^{-\rho(\tau+(s-t_0))}(X_\tau^x- a)]$$

$$= e^{\rho t_0} \sup_\tau E[e^{-\tau(t+s)}(X_\tau^x- a)] = e^{\rho t_0} g^*(s,x).$$

Therefore the connected component of D that contains U must have the form

$$D(x_0) = \{(t,x);\ x<x_0\} \quad \text{for some } x_0> \frac{a\rho}{\rho-r}.$$

Note that D cannot have any other components, for if V is a component of D disjoint from U then $\mathcal{A}g < 0$ in V and so

$$E^y[g(Y_\tau)] = g(y) + E^y[\int_0^\tau \mathcal{A}g(Y_s)ds] < g(y)$$

for all exit times τ bounded by the exit time from an x-bounded strip in V. From this we conclude by Theorem 10.9 c) that $g^*= g$, which implies $V = \emptyset$.

Put $\tau(x_0) = \tau_{D(x_0)}$ and let us compute

(10.38) $\tilde{g}(t,x) = \tilde{g}_{x_0}(t,x) = E^{(t,x)}[g(Y_{\tau(x_0)})]$

From Chapter IX we know that $f = \tilde{g}$ is the solution of the boundary value problem

(10.39) $$\frac{\partial f}{\partial t} + rx\,\frac{\partial f}{\partial x} + \frac{1}{2}\,\alpha^2x^2\,\frac{\partial^2 f}{\partial x^2} = 0 \quad \text{for } x < x_0$$

$$f(t,x_0) = e^{-\rho t}(x_0- a)$$

(Note that $\mathbb{R} \times \{0\}$ does not contain any regular boundary points of D wrt. $Y_s= (t+s, X_s)$)

If we try a solution of (10.39) of the form

$$f(t,x) = e^{-\rho t}\phi(x)$$

we get the following 1-dimensional problem

$$-\rho\phi + rx\phi'(x) + \tfrac{1}{2}\alpha^2x^2\phi''(x) = 0 \quad \text{for } x < x_0$$
$$\phi(x_0) = x_0 - a$$

The substitution $x = e^y$ transforms the equation into an equation with constant coefficients which we can solve explicity. We find

$$\phi(x) = C_1x^{\gamma_1} + C_2x^{\gamma_2},$$

where

$$\gamma_i = \alpha^{-2}\left[\tfrac{1}{2}\alpha^2 - r \pm \sqrt{(r-\tfrac{1}{2}\alpha^2)^2 + 2\rho\alpha^2}\,\right], \quad \gamma_2 < 0 < \gamma_1$$

Since $\phi(x)$ is bounded as $x \to 0$ we must have $C_2 = 0$ and the boundary requirement $\phi(x_0) = x_0 - a$ gives $C_1 = x_0^{-\gamma_1}(x_0 - a)$. We conclude that the solution f of (10.38) is

$$(10.40) \qquad \tilde{g}_{x_0}(t,x) = f(t,x) = e^{-\rho t}(x_0 - a)\left(\frac{x}{x_0}\right)^{\gamma_1}$$

If we fix (t,x) then the value of x_0 which maximizes $\tilde{g}_{x_0}(t,x)$ is easily seen to be given by

$$(10.41) \qquad x_0 = x_{max} = \frac{a\gamma_1}{\gamma_1 - 1}$$

(Note that $\gamma_1 > 1$ if and only if $r < \rho$)

So by (10.38-41) we can conclude that

$$g^*(t,x) = \sup_\tau E^{(t,x)}[g(\tau, X_\tau)] = \sup_{x_0} E^{(t,x)}[g(\tau(x_0), X_{\tau(x_0)})]$$

$$= \sup_{x_0} \tilde{g}_{x_0}(t,x) = \tilde{g}_{x_{max}}(t,x).$$

The conclusion is therefore that one should sell the assets the first time the price of them reaches the value $x_{max} = \frac{a\gamma_1}{\gamma_1 - 1}$. The expected discounted profit obtained from this strategy is

$$g^*(t,x) = e^{-\rho t}\left(\frac{\gamma_1 - 1}{a}\right)^{\gamma_1 - 1}\left(\frac{x}{\gamma_1}\right)^{\gamma_1}.$$

<u>REMARK</u>. The reader is invited to check that the value $x_0 = x_{max}$ is the only value of x_0 which makes the function

$$x \to \tilde{g}_{x_0}(t,x) \qquad \text{(given by (10.38))}$$

continuously differentiable at x_0. This is not a coincidence. In fact, it illustrates a general phenomenon which is known as the <u>high contact</u> (or smooth fit) <u>principle</u>. See Samuelson [1], McKean [1], Bather [1] and Shiryayev [1]. This principle is the basis of the fundamental connection between optimal stopping and <u>free boundary value problems</u>. More information can be found in Bensoussan and Lions [1] and Friedman [2].

After these continuous-time-and-space examples, we would like to end this chapter with two 'discrete' examples. Although these two specific examples may of course be solved directly without the use of Ito diffusions, they may serve to illustrate the content and power of the general solutions we have found.

<u>EXAMPLE 10.17</u>. A participant in a contest is given a sequence of questions. If she gives the right answer to a question, she gets a reward plus the option of proceeding to the next question or withdraw from the contest (with the money she has received so far). If she cannot answer a question correctly she looses all the money she has received and she is out of the contest. Suppose that for each question there is a probability p that she can answer the question. When is the best time for the contestant to stop the game? What is the expected gain when stopping at such an optimal time? Strictly speaking the situation in this problem is not covered by our continuous-time-approach, but it is not hard to see that with the proper modifications our proofs carry over to the analogues discrete situation. This example may serve as an illustration of what the corresponding discrete notions are:

We let X_n denote the state of the contest at time $t = n$ $(n = 1,2,\ldots)$. The state 0 designates that a wrong answer has been given and that the contest is permanently over. The state $k \geq 1$ designates that k consecutive right answers have been given. Let p denote the probability that the contestant answers a question correctly. Then the probability that the state changes from i to j during one question, P_{ij}, is given by the following infinite matrix $P = [P_{ij}]$:

$$P = \begin{array}{c|cccccc} & \text{state } 0 & \text{state } 1 & \text{state } 2 & \cdots & & \\ \hline \text{state } 0 & 1 & 0 & 0 & \cdots & & \\ \text{state } 1 & 1-p & 0 & p & \cdots & & \\ \text{state } 2 & 1-p & 0 & 0 & p & & \\ \cdot & 1-p & 0 & 0 & 0 & p & \\ \cdot & \cdot & \cdot & \cdot & & & \ddots \\ \cdot & \cdot & \cdot & \cdot & & & \end{array}$$

Thus $\{X_n\}$ is a Markov chain with transition matrix P and state space $\{0,1,2,\ldots\} = S$. A function $f: S \to R$, i.e. an infinite vector $f = (f_j)_{j=0}^{\infty}$, is called superharmonic (wrt. X_n) if

$$f_i \geq \sum_{j=0}^{\infty} P_{ij} f_j \quad \text{for all } i \in S.$$

In this example the reward function $g = (g_j)$ is given by

$$g_j = j \cdot a \; ; \quad j = 0, 1, 2, \ldots$$

where a is the reward obtained for each correct answer. Theorem 10.9 tells us that the optimal reward g^* - if it exists - has the form

$$g_j^{(m)} = \begin{cases} \tilde{g}_j & \text{for } j < m \\ g_j & \text{for } j \geq m \end{cases}$$

for some m, where

$$\tilde{g}_j = g_m \cdot P[X_\tau = m | X_0 = j] = g_m \cdot p^{m-j} \; ; \; 1 \leq j \leq m-1,$$

and $\tau = \min\{k>0 \; ; \; X_k = 0 \text{ or } X_k = m\}$ (i.e. $\tilde{g}$ is the "harmonic extension" of $g|\{0,m\}$ to $\{1,2,\ldots,m-1\}$).
Thus

$$g_j^{(m)} = \begin{cases} 0 & \text{if } j = 0 \\ map^{m-j} & \text{if } 1 \leq j \leq m-1. \\ ja & \text{if } j \geq m \end{cases}$$

Since

$$\sum_j P_{ij} g_j^{(m)} = \begin{cases} 0 & \text{if } i = 0 \\ map^{m-i} & \text{if } 1 \leq i \leq m-2 \\ (i+1)ap & \text{if } i \geq m-1, \end{cases}$$

$g^{(m)}$ is superharmonic iff

$$(i+1)ap \leq ia \quad \text{for all } i \geq m$$

i.e. iff

$$(m+1)p \leq m$$

i.e. iff

$$m \geq \frac{p}{1-p} .$$

Choose m_0 to be the smallest integer $\geq \frac{p}{1-p}$. Then we claim that $g^{(m_0)}$ majorizes g. To obtain this it is enough to prove that

$$m_0 a p^{m_0 - j} \geq ja \quad \text{for} \quad 1 \leq j \leq m_0$$

i.e. $$m_0 p^{m_0} \geq j p^j \quad \text{for} \quad 1 \leq j \leq m_0 .$$

Now

$$(j+1)p^{j+1} - jp^j = (1+j)p^j\left(p - \frac{j}{1+j}\right) \geq 0 \quad \text{for} \quad 1 \leq j \leq m_0 ,$$

since

$$p \geq \frac{m_0 - 1}{m_0} \geq \frac{j}{1+j} \quad ; \; 1 \leq j \leq m_0 - 1.$$

Thus $g^{(m_0)}$ is a superharmonic majorant of g and therefore $g^* = g^{(m_0)}$ by Corollary 9.10, and the continuation region is $D = \{1,2,\ldots, m_0 - 1\}$.

We conclude that the optimal reward is

$$g_j^* = g_j^{(m_0)} = \begin{cases} 0 & \text{if} \quad j = 0 \\ m_0 a p^{m_0 - j} & \text{if} \quad 1 \leq j \leq m_0 - 1, \\ ja & \text{if} \quad j \geq m_0 \end{cases}$$

(thus the optimal reward is $m_0 a p^{m_0 - 1}$ if the contest starts in state 1) and the optimal stopping rule is to stop the game the first time X_j is in the state m_0, i.e. the first time the contestant has obtained at least $\frac{p}{1-p}$ consecutive right answers.

EXAMPLE 10.18. (The optimal selection problem, also called the secretary problem or the marriage problem).
At the times $t_k = k$ $(k=1,2,\ldots,n)$ the objects a_k passes (in a random order) for inspection, one at a time. Each object can be evaluated and will be either inferior or superior to any of the other objects. Every time an object passes for inspection one has the choice of either accepting the object (and the process stops) or rejecting the object (for good) and the next appears. What selection strategy should one use to maximize the probability of obtaining the best object?

If T_0 is given, define inductively T_{i+1} to be the first time after T_i when an object appears that is superior to the object a_{T_i}. $T_{i+1} = \infty$ if no object appears which is superior to a_{T_i}. In order words, T_i is the time of the i'th record. Then $\{T_i\}$ is a Markov chain with transition function

$$(10.42) \qquad P_{k,\ell} = P[T_{i+1}=\ell \mid T_i=k] = \begin{cases} \frac{k}{\ell(\ell-1)} & \text{for } k < \ell \leq n \\ 0 & \text{for } k \geq \ell \\ \frac{k}{n} & \text{for } \ell = \infty \end{cases}$$

This is because $P[T_{i+1}= \ell \mid T_i=k]$ is the probability that the next record occurs at time ℓ, given that the present record occurred at time k (which is clearly independent of i, since the order of the first k objects is arbitrary), and this probability is

$$\frac{P[\text{present record occurred at time } k \text{ \& next record at time } \ell]}{P[\text{present record occurred at time } k]}$$

Now

P[present record occurred at time k] =

P[the best of the first k objects came at time k] $= \frac{1}{k}$ and (if $\ell<\infty$)

P[present record occurred at time k & next record at time ℓ]

= P[the best of ℓ objects came at ℓ, the next best at k]

$= \frac{(\ell-2)!}{\ell!} = \frac{1}{\ell(\ell-1)}$, and this establishes (10.42).

The reward function.

Suppose we stop the process T_i when $T_i= x$. Then the i'th record has ocurred at time x. The probability that this i'th record is the absolute maximum is the same as the probability that the maximum of the first x objects is the absoulute maximum, i.e. $\frac{x}{n}$. So the reward function is

$$(10.43) \qquad g(x) = \frac{x}{n} \quad \text{for } x \leq n \quad (g(\infty)=0)$$

Therefore the problem is to find a stopping time τ^* such that

$$g^*(1) = E^1[g(T_{\tau^*})] = \sup_{\tau} E^1[g(T_\tau)]$$

Now

$$(Pg)(k) = \sum_{\ell=k+1}^{n} P_{k,\ell}\, g(\ell) = \sum_{\ell=k+1}^{n} \frac{k}{\ell(\ell-1)} \cdot \frac{\ell}{n} = \frac{k}{n}\left[\frac{1}{k} + \frac{1}{k+1} + \cdots + \frac{1}{n-1}\right]$$

so $(Pg)(k) > g(k)$ if $\frac{1}{k} + \frac{1}{k+1} + \cdots + \frac{1}{n-1} > 1$, i.e. if $k \leq m$,

where m is the largest integer such that

$$\frac{1}{m} + \frac{1}{m+1} + \cdots + \frac{1}{n-1} > 1$$

Thus $\hat{g}(k) \geq (P\hat{g})(k) \geq (Pg)(k) > g(k)$ if $k \leq m$,
i.e.

$$\{1,2,\ldots,m\} \subset D = \{k;\ g(k) < \hat{g}(k)\},\quad \text{the continuation region.}$$

Define

$$g^{(m)}(x) = \begin{cases} \tilde{g}(x) & \text{if } x \leq m \\ g(x) & \text{if } x > m, \end{cases}$$

where

$$\tilde{g}(x) = \sum_{k=m+1}^{n} g(k)\ P^x[T_\tau = k] \quad \text{is the value at } x$$

of the harmonic extension of $g|\{m+1,\ldots\}$ to $\{1,2,\ldots,m\}$, so that τ is the first exit time of T_i from $\{1,2,\ldots,m\}$.
Since

$$P^x[T_\tau = k] = P^x[\text{first record after } m \text{ occurs at time } k] = \frac{m}{k(k-1)}$$

if $x \leq m$, we have

$$\tilde{g}(x) = \sum_{k=m+1}^{n} \frac{k}{n}\cdot\frac{m}{k(k-1)} = \frac{m}{n}\left[\frac{1}{m} + \cdots + \frac{1}{n-1}\right],\quad x \leq m.$$

Hence $(Pg^{(m)})(k) = \frac{k}{n}\left[\frac{1}{m} + \cdots + \frac{1}{n-1}\right]\left[1 - \frac{k}{m}\right] \leq \tilde{g}(k)$ if $k \leq m$.

If $k > m$ then $(Pg^{(m)})(k) = \frac{k}{n}\left[\sum_{k+1}^{n} \frac{1}{\ell-1}\right] \leq g^{(m)}(k)$.

Finally, we have

$$g^{(m)}(x) = g(x) \quad \text{for } x > m$$

and

$$g^{(m)}(x) = \frac{m}{n}\left[\frac{1}{m} + \ldots + \frac{1}{n-1}\right] > \frac{x}{n} = g(x) \quad \text{if } x \leq m.$$

Thus $g^{(m)}$ is a superharmonic majorant of g, so by Corollary 10.10 we have

$$D = \{1,2,\ldots,m\}$$

and $\tau = \tau_D = \inf\{t>0;\ T_t > m\}$ is optimal.
Thus the conclusion is that one should stop at the first time a new record appears after time m. Note that

$$m \sim \frac{n}{e}$$

and that

$$g^*(1) = \frac{m}{n}\left[\frac{1}{m} + \cdots + \frac{1}{n-1}\right] \sim \frac{1}{e} = 0{,}368\cdots \quad \text{for large } n.$$

So, perhaps surprisingly, no matter how large n is one obtains with this strategy the best object with probability no less than 0.368 $\cdots$.

XI. Application to Stochastic Control

Statement of the problem. Suppose that the state of a system at time t is described by a stochastic integral X_t of the form

$$(11.1)\qquad dX_t = dX_t^u = b(t,X_t,u)dt + \sigma(t,X_t,u)dB_t \ ,$$

where $X_t, b \in \mathbb{R}^n$, $\sigma \in \mathbb{R}^{n\times m}$ and B_t is m-dimensional Brownian motion. Here $u \in R^k$ is a parameter whose value we can choose at any instant in order to control the process X_t. Thus $u=u(t,\omega)$ is a stochastic process. Since our decision at time t must be based upon what has happened up to time t, the function $\omega\to u(t,\omega)$ must (at least) be measurable wrt. $\mathcal{F}_t$, i.e. the process u_t must be $\mathcal{F}_t$-adapted. Thus the right hand side of (11.1) is well-defined as a stochastic integral, under suitable assumptions on b and σ. At the moment we will not specify the conditions on b and σ further, but simply assume that the process X_t satisfying (11.1) exists. See further comments on this in the end of this chapter.

Let $\{X_h^{t,x}\}_{h\geq t}$ be the solution of (11.1) such that $X_t^{t,x}=x$, i.e. $X_h^{t,x} = x + \int_t^h b(r,X_r^{t,x},u)dr + \int_t^h \sigma(r,X_r^{t,x},u)dB_r$ and let the probability law of X_s starting at x for $s=t$ be denoted by $Q^{t,x}$, so that

$$(11.2)\qquad Q^{t,x}[X_{t_1}\in E_1,\dots,X_{t_k}\in E_k] = P^0[X_{t_1}^{t,x}\in E_1,\dots,X_{t_k}^{t,x}\in E_K]$$

for $t\leq t_i;\ 1\leq i\leq k$.

To obtain an easier notation we introduce

$$Y_s = (t+s, X_{t+s}^{t,x}) \quad \text{for} \quad s\geq 0, \quad Y_0=(t,x)$$

and we observe that if we substitute in (11.1) we get the equation

$$(11.3)\qquad dY_t = dY_t^u = b(Y_t,u)dt + \sigma(Y_t,u)dB_t \ ,$$

i.e. the stochastic integral is time-homogeneous. (Strictly speaking, the u,b and σ in (11.3) are slightly different from the u,b and σ in (11.1).) The probability law of Y_s starting at $y=(t,x)$ for $s=0$ is (with abuse of notation) also denoted by $Q^{t,x}=Q^y$.

We assume that a cost function (or performance criterion) has been given on the form

(11.4) $$J(t,x,u) = E^{t,x}\left[\int_t^\tau F(s,X_s,u)ds + K(\tau,X_\tau)\right]$$

or with $y=(t,x)$,

$$J^u(y) = E^y\left[\int_0^\tau F^u(Y_s)ds + K(Y_\tau)\right], \quad \text{with} \quad F^u(y)=F(y,u),\ J^u(y)=J(Y,u),$$

where K is a bounded "bequest" function, F is bounded and continuous and τ is assumed to be the exit time of Y_s from some (fixed) open set $G \subset \mathbb{R}^{n+1}$. Thus, in particular, τ could be a fixed time t_0. We assume that $E^y[\tau]<\infty$ for all $y\in G$.

The problem is to find a control function $u^*=u^*(t,\omega)$ such that

(11.5) $$H(y) \overset{\text{def}}{=} \inf_{u(t,\omega)} \{J^u(y)\} = J^{u^*}(y) \quad \text{for all} \quad y\in G$$

where the inf is taken over all $\mathcal{F}_t$-adapted processes $\{u_t\}$, usually required to satisfy some extra conditions. Such a control u^* - if it exists - is called an <u>optimal control</u>. Examples of types of control functions that may be considered:

1) Functions $u(t,\omega)=u(t)$ i.e. not depending on ω. These controls are sometimes called <u>deterministic</u> or <u>open loop controls</u>.
2) Processes $\{u_t\}$ which are $\mathcal{M}_t$-adapted, i.e. for each t the function $\omega\to u(t,\omega)$ is $\mathcal{M}_t$-measurable, where $\mathcal{M}_t$ is the σ-algebra generated by $\{X_s^u;\ s\leq t\}$. These controls are called <u>closed loop</u> or <u>feedback</u> controls.
3) The controller has only <u>partial knowledge</u> of the state of the system. More precisely, to the controller's disposal are only (noisy) observations R_t of X_t, given by a stochastic integral of the form

$$dR_t = a(t,X_t)dt + \gamma(t,X_t)dB .$$

Hence the control process $\{u_t\}$ must be adapted wrt. the σ-algebra $\mathcal{N}_t$ generated by $\{R_s;\ s\leq t\}$. In this situation the stochastic control problem is linked to the filtering problem (Chapter VI). In fact, if the equation (11.1) is linear and the cost function is integral quadratic (i.e. F and K are quadratic) then the stochastic control problem splits into a linear filtering problem and a corresponding deterministic control problem. This is called The Separation Principle. See Example 11.4.

4) Functions $u(t,\omega)$ of the form $u(t,\omega)=u_0(t,X_t(\omega))$ for some function $u_0\colon \mathbb{R}^{n+1} \to \mathbb{R}^k$: These are called Markov controls, because with such u the corresponding process X_t becomes an Ito diffusion, in particular a Markov process. (For simplicity we will assume that u satisfies (5.14) and (5.15).) In the following we will not distinguish between u and u_0. Thus we will identify a function $u\colon \mathbb{R}^{n+1} \to \mathbb{R}^k$ with the Markov control $u(Y)=u(t,X_t)$ and simply call such functions Markov controls.

The Hamilton-Jacobi-Bellman equation

Let us first consider only Markov controls

$$u = u(t,X_t(\omega))$$

Introducing $Y_s=(t+s,X_{t+s})$ (as explained earlier) the system equation becomes

$$dY_t = b(Y_t,u(Y_t))dt + \sigma(Y_t,u(Y_t))dB_t \,. \tag{11.6}$$

For $v \in \mathbb{R}^k$ define

$$(A^v f)(y) = \frac{\partial f}{\partial t}(y) + \sum b_i(y,v)\frac{\partial f}{\partial x_i} + \sum a_{ij}(y,v)\frac{\partial x^2 f}{\partial x_i \partial x_j} \tag{11.7}$$

where $a_{ij} = \frac{1}{2}(\sigma\sigma^T)_{ij}$, $y=(t,x)$ and $x=(x_1,\ldots,x_n)$. Then for each choice of the function u the solution $Y_t=Y_t^u$ is an Ito diffusion with generator

$$(Af)(y) = (A^{u(y)}f)(y) \qquad \text{(see Theorem 7.9)}.$$

For $v \in \mathbb{R}^k$ define $F^v(y)=F(y,v)$. The fundamental result in stochastic control theory is the following:

Theorem 11.1. (The Hamilton-Jacobi-Bellman (HJB) equation)

Define

$$H(y) = \inf\{J^u(y);\ u=u(Y) \text{ Markov control}\}$$

Suppose that $H \in C^2$ and that an optimal Markov control u^* exists. Then

$$\inf_{v\in\mathbb{R}^k}\{F^v(y)+(A^v H)(y)\} = 0 \quad \text{for all } y\in G \tag{11.8}$$

and

$$H(y) = K(y) \quad \text{for all } y\in\partial_R G \,. \tag{11.9}$$

where $\partial_R G$ is the set of regular boundary points of G for $Y_t^{u^*}$. The infimum in (11.8) is obtained if $v=u^*(y)$ where $u^*(y)$ is optimal. In other words,

$$F(y,u^*(y)) + (A^{u^*}H)(y) = 0 \quad \text{for all } y\in G . \tag{11.10}$$

<u>Proof</u>: The last two statements are easy to prove:
Since $u^*=u^*(y)$ is optimal we have

$$H(y) = J^{u^*}(y) = E^y[\int_0^\tau F(Y_s,u^*(Y_s))ds+K(Y_\tau)]$$

If $y\in\partial_R G$ then $\tau=0$ a.s. Q^y and (11.9) follows. By the solution of the Dirichlet-Poisson problem (Theorem 9.18)

$$(A^{u^*(y)}H)(y) = -F(y,u^*(y)) \quad \text{for all } y\in G ,$$

which is (11.10). We proceed to prove (11.8). Fix $y=(t,x)\in G$. Let $\alpha<\tau$ be some bounded stopping time. Then we obtain from the Dynkin formula

$$E^y[H(Y_\alpha)] = H(y) + E^y[\int_0^\alpha (A^u H)(Y_r)dr] . \tag{11.11}$$

Since

$$J^u(y) = E^y[\int_0^\tau F^u(Y_r)dr+K(Y_\tau)] ,$$

we get by the strong Markov property, (7.16) and (9.27)

$$\begin{aligned} E^y[J^u(Y_\alpha)] &= E^y[E^{Y_\alpha}[\int_0^\tau F^u(Y_r)dr+K(Y_\tau)]] \\ &= E^y[E^y[\theta_\alpha(\int_0^\tau F^u(Y_r)dr+K(Y_\tau))\mid \mathcal{F}_\alpha]] \\ &= E^y[E^y[\int_\alpha^\tau F^u(Y_r)dr+K(Y_\tau)\mid \mathcal{F}_\alpha]] \\ &= E^y[\int_0^\tau F^u(Y_r)dr+K(Y_\tau)-\int_0^\alpha F^u(Y_r)dr] \\ &= J^u(y) - E^y[\int_0^\alpha F^u(Y_r)dr] . \end{aligned}$$

So

$$J^u(y) = E^y[\int_0^\alpha F^u(Y_r)dr)]+E^y[J^u(Y_\alpha)] \tag{11.12}$$

This holds for all Markov controls u.

Now let $U \subset G$ be of the form $U=\{(r,z)\in G;\ r<t_1\}$ where $t_1>t$ and put $\alpha=\tau_U$, the first exit time from U of Y_s. Suppose an optimal control $u^*(y)=u^*(r,z)$ exists and put

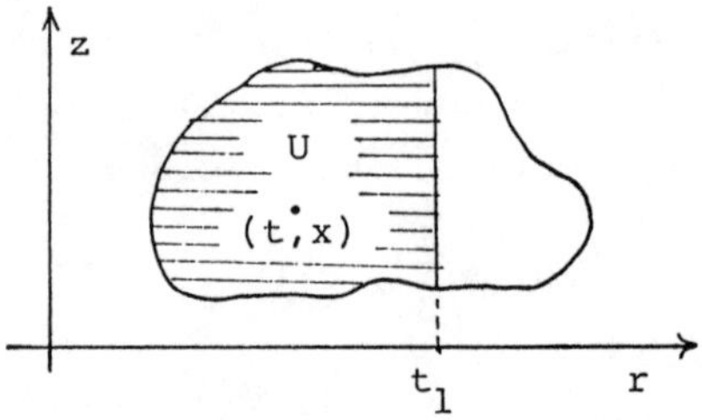

$$u(r,z) = \begin{cases} v & \text{if } (r,z)\in U \\ u^*(r,z) & \text{if } (r,z)\in G\setminus U \end{cases}$$

where $v \in \mathbb{R}^k$ is arbitrary. Then

$$H(Y_\alpha) = J^{u^*}(Y_\alpha) = J^u(Y_\alpha) \tag{11.13}$$

and therefore, combining (11.11) and (11.12) we obtain (approximating α by bounded α_k we see that (11.12) remains true for our choice of α)

$$H(y) \le J^u(y) = E^y[\int_0^\alpha F^v(Y_r)dr] + E^y[H(Y_\alpha)]$$

$$= E^y[\int_0^\alpha F^v(Y_r)dr] + H(y) + E^y[\int_0^\alpha (A^vH)(Y_r)dr]$$

or

$$E^y[\int_0^\alpha (F^v(Y_r)+(A^vH)(Y_r))dr] \ge 0$$

So

$$\frac{E^y[\int_0^\alpha (F^v(Y_r)+(A^vH)(Y_r))dr]}{E^y[\alpha]} \ge 0 \qquad \text{for all such } U.$$

Letting $t_1 \downarrow t$ we obtain, since $F^v(\cdot)$ and $(A^vH)(\cdot)$ are continuous at y, that $F^v(y)+(A^vH)(y)\ge 0$, which combined with (11.10) gives (11.8). That completes the proof.

<u>REMARK</u>. The HJB equation states that if an optimal control u^* exists, then we know that its value v at the point y is a point v where the function

$$v \to F^v(y) + (A^vH)(y)$$

attains its minimum (and this minimum is 0). Thus the original stochastic control problem is associated to the easier problem of finding the minimum of a real function in $\mathbb{R}^k$. However, the HJB equation only states that it is <u>necessary</u> that $v=u^*(y)$ is the minimum of this function. It is just as important to know if this is also sufficient: If at each point y we have found $v=u_0(y)$ such that $F^v(y)+(A^vH)(y)$ is minimal, will $u_0(Y)$ be an optimal control? The next result states that (under some conditions) this is actually the case:

THEOREM 11.2. (A converse of the HJB equation)
Let h be a bounded function in $C^2(G)\cap C(\bar{G})$ such that, for all $v \in \mathbb{R}^k$,

$$F^v(y) + (A^v h)(y) \geq 0 \; ; \quad y\in G \tag{11.14}$$

with boundary values

$$h(y) = K(y) \quad ; \quad y\in\partial G \; . \tag{11.15}$$

Then

$$h(y) \leq J^u(y) \quad \text{for all Markov controls } u \text{ and all } y\in G. \tag{11.16}$$

Moreover, if for each $y\in G$ we have found $u_0(y)$ such that

$$F^{u_0(y)}(y) + (A^{u_0(y)}h)(y) = 0 \tag{11.17}$$

then $u_0=u_0(Y)$ is a Markov control such that

$$h(y) = J^{u_0(y)}(y) \tag{11.18}$$

and hence u_0 must be an optimal control (by (11.16)).

Proof. Assume that h satisfies (11.14) and (11.15) above. Let u be a Markov control. Since $A^u h \geq -F^u$ in G we have by Dynkin's formula

$$\begin{aligned} E^y[h(Y_{\tau\wedge k})] &= h(y) + E^y[\int_0^{\tau\wedge k}(A^u h)(Y_r)dr] \\ &\geq h(y) - E^y[\int_0^{\tau\wedge k}F^u(Y_r)dr] \end{aligned}$$

for all integers k, such that

$$h(y) \leq E^y[\int_0^\tau F^u(Y_r)dr+h(Y_\tau)] = E^y[\int_0^\tau F^u(Y_r)dr+K(Y_\tau)] = J^u(y) \; ,$$

which proves (11.16). If u_0 is such that (11.17) holds, then the calculations above give equality and the proof is complete.

The HJB equation and its converse provide a very nice solution to the stochastic control problem in the case where only Markov controls are considered. One might feel that considering only Markov controls is too restrictive, but fortunately one can always obtain as good performance with a Markov control as with an arbitrary $\mathcal{F}_t$-adapted control, at least if some extra conditions are satisfied:

THEOREM 11.3. Let

$$H_M(y) = \inf\{J^u(y);\ u=u(Y)\ \ \text{Markov control}\}$$

and

$$H_a(y) = \inf\{J^u(y);\ u=u(t,\omega)\ \ \mathcal{F}_t\text{-adapted control}\}\ .$$

Suppose there exists an optimal Markov control $u_0=u_0(Y)$ for the Markov control problem (i.e. $H_M(y)=J^{u_0}(y)$ for all $y\in G$) such that all the boundary points of G are regular wrt. $Y_t^{u_0}$ and that H_M is a bounded function in $C^2(G)\cap C(\bar{G})$.
Then

$$H_M(y) = H_a(y) \quad \text{for all } y\in G\ .$$

Proof. Let h be a bounded function in $C^2(G)\cap C(\bar{G})$ satisfying

$$F^v(y) + (A^v h)(y) \geq 0 \quad \text{for all } y\in G,\ v\in \mathbb{R}^k \tag{11.19}$$

and

$$h(y) = K(y) \qquad\qquad \text{for all } y\in\partial G\ . \tag{11.20}$$

Let $u_t(\omega)=u(t,\omega)$ be an $\mathcal{F}_t$-adapted control. Then Y_t is a stochastic integral given by

$$dY_t = b(Y_t,u_t)dt + \sigma(Y_t,u_t)dB_t$$

so by Lemma 7.8

$$E^y[h(Y_\tau)] = h(y) + E^y[\int_0^\tau (A^{u(s,\omega)}h)(Y_s)ds], \tag{11.21}$$

where $(A^{u(s,\omega)}h)(y) = \frac{\partial h}{\partial t}(y) + \sum b_i(y,u(s,\omega))\frac{\partial h}{\partial x_i}(y) + \sum a_{ij}(y,u(s,\omega))\frac{\partial^2 h}{\partial x_i\partial x_j}(y)$, with $a_{ij} = \frac{1}{2}(\sigma\sigma^T)_{ij}$. Thus by (11.19) and (11.20) this gives

$$E^y[K(Y_\tau)] \geq h(y) - E^y[\int_0^\tau F(Y_s,u(s,\omega))ds] \tag{11.22}$$

or

$$h(y) \leq J^u(y)\ . \tag{11.23}$$

But by Theorem 11.1 the function $h(y)=H_M(y)$ satisfies (11.19) and (11.20). So by (11.23) we have $H_M(y) \leq H_a(y)$ and Theorem 11.3 follows.

<u>REMARK</u>. In the proof above Lemma 7.8 does not apply to h directly, since h need not have compact support. However, we can still obtain (11.22) by first applying Lemma 7.8 to stopping times of the form $\tau_r=\tau\wedge\sigma_r$, where $\sigma_r=\inf\{s>0;\ |Y_s|>r\}$ (i.e. redefining h outside the ball of radius r so that it has compact support) and then letting $r\to\infty$ in the corresponding (11.22)-formula for τ_r.

We now illustrate the results by looking at some examples:

<u>EXAMPLE 11.4</u>. (The linear regulator problem).

Suppose that the state X_t of the system at time t is given by a linear stochastic differential equation:

(11.24) $$dX_t = (L_tX_t+M_tu_t)dt + \sigma_t dB_t \ ,$$

and the cost is of the form

(11.25) $$J^u(t,x) = E^{t,x}\Big[\int_0^\tau\{X_s^TC_sX_s+u_s^TD_su_s\}ds+X_\tau^TRX_\tau\Big] \ ,$$

where all the coefficients $L_t\in\mathbb{R}^{n\times n}$, $M_t\in\mathbb{R}^{n\times k}$, $\sigma_t\in\mathbb{R}^{n\times m}$, $C_t\in\mathbb{R}^{n\times n}$, $D_t\in\mathbb{R}^{k\times k}$ and $R\in\mathbb{R}^{n\times n}$ are t-continuous, nonrandom. We assume that C_t and R are symmetric, non-negative definite and D_t is symmetric, positive definite, for all t. We will also assume that τ is a non-random time t_1.

The problem is then to choose the control $u=u(t,X_t)$ such that it minimizes $J^u(t,x)$. We may interpret this as follows: The aim is to find a control u which makes $|X_t|$ small fast and such that the energy used ($\sim u^TDu$) is small. The sizes of C_t and R reflect the cost of having large values of $|X_t|$, while the size of D_t reflects the cost (energy) of applying large values of $|u_t|$.

In this case the HJB-equation becomes

(11.26) $$0 = \inf_v\{F^v(t,x)+(A^vH)(t,x)\} =$$
$$\frac{\partial H}{\partial t} + \inf_v\{x^TC_tx+v^TD_tv+\sum_i(L_tx+M_tv)_i\frac{\partial H}{\partial x_i} + \tfrac{1}{2}\sum(\sigma\sigma^T)_{ij}\frac{\partial^2H}{\partial x_i\partial x_j}\}$$

and

(11.27) $$H(t_1,x) = x^TRx \ .$$

The problem with this equation is that we don't know what H is. Let us try a function of this form

$$h(t,x) = x^T S_t x + a_t \tag{11.28}$$

where $S(t) = S_t \in \mathbb{R}^{n\times n}$ is symmetric, non-negative definite, $a_t \in \mathbb{R}$ and both a_t and S_t are continuously differentiable wrt. t (non-random). In order to use Theorem 11.2 we need to determine S_t and a_t such that

$$h(t_1,x) = x^T R x \tag{11.30}$$

To obtain (11.30) we put

$$S_{t_1} = R \tag{11.31}$$

$$a_{t_1} = 0 \tag{11.32}$$

Using (11.28) we get

$$F^v(t,x) + (A^v h)(t,x) = x^T \dot{S}_t x + \dot{a}_t + x^T C_t x + v^T D_t v + \\ + (L_t x + M_t v)^T (S_t x + S_t^T x) + 2\cdot\tfrac{1}{2}\sum_{i,j}(\sigma\sigma^T)_{ij} S_{ij}, \tag{11.33}$$

where $\dot{S}_t = \frac{d}{dt}S_t$. The minimum of this expression is obtained when

$$\frac{\partial}{\partial v_i}(F^v(t,x)+(A^v h)(t,x)) = 0 \;; \quad i=1,\ldots,k$$

i.e. when

$$2D_t v + 2M_t^T S_t x = 0$$

i.e. when

$$v = -D_t^{-1} M_t^T S_t x \;. \tag{11.34}$$

We substitute this value of v in (11.33) and obtain

$$\begin{aligned} F^v(t,x) + (A^v h)(t,x) &= \\ &= x^T \dot{S}_t x + \dot{a}_t + x^T C_t x + x^T S_t M_t D_t^{-1} D_t D_t^{-1} M_t^T S_t x \\ &+ (L_t x - M_t D_t^{-1} M_t^T S_t x)^T 2 S_t x + \operatorname{tr}(\sigma\sigma^T S)_t \\ &= x^T(\dot{S}_t + C_t - S_t M_t D_t^{-1} M_t^T S_t + 2L_t^T S_t)x + \dot{a}_t + \operatorname{tr}(\sigma\sigma^T S)_t \;, \end{aligned}$$

where tr denotes the (matrix) trace.

We obtain that this is 0 if we choose S_t such that

$$\dot{S}_t = -2L_t^T S_t + S_t M_t D_t^{-1} M_t^T S_t - C_t \quad ; \quad t<t_1 \tag{11.35}$$

and a_t such that

$$\dot{a}_t = -\mathrm{tr}(\sigma\sigma^T S)_t \quad ; \quad t<t_1 . \tag{11.36}$$

We recognize (11.35) as a Riccati type equation from linear filtering theory (see (6.42)). Equation (11.35) with boundary condition (11.31) determines S_t uniquely. Combining (11.36) with the boundary condition (11.32) we obtain

$$a_t = \int_t^{t_1} \mathrm{tr}(\sigma\sigma^T S)_s ds \tag{11.37}$$

With such a choice of S_t and a_t we see that (11.29) and (11.30) hold, so by Theorem 11.2 we conclude that

$$u^*(t,x) = -D_t^{-1} M_t^T S_t x , \qquad t<t_1 \tag{11.38}$$

is an optimal control and the minimum cost is

$$H(t,x) = x^T S_t x + \int_t^{t_1} \mathrm{tr}(\sigma\sigma^T S)_s ds , \quad t<t_1 . \tag{11.39}$$

This formula shows that the extra cost due to the noise in the system is given by

$$a_t = \int_t^{t_1} \mathrm{tr}(\sigma\sigma^T S)_s ds .$$

The Separation Principle (see Davis [1] or Fleming and Rishel [1]) states that if we had only partial knowledge of the state X_t of the system, i.e. if we only had noisy observations

$$dZ_t = g_t X_t dt + \gamma_t d\tilde{B}_t \tag{11.40}$$

to our disposal, then the optimal control $u^*(t,\omega)$ (required to be $\mathcal{G}_t$-adapted, where $\mathcal{G}_t$ is the σ-algebra generated by $\{Z_s; s\leq t\}$), would be given by

$$u^*(t,\omega) = -D_t^{-1} M_t^T S_t \hat{X}_t(\omega) , \tag{11.41}$$

where $\hat{X}_t$ is the filtered estimate of X_t based on the observations $\{Z_s; s\leq t\}$, given by the Kalman-Bucy filter (6.44). Comparing with (11.38) we see that the stochastic control problem in this case splits into a linear filtering problem and a deterministic control problem.

An important field of applications of the stochastic control theory is economics and finance. Therefore we illustrate ths results above by applying them to a simple case of optimal portfolio diversification. This problem has been considered in more general settings by many authors, see for example Markowitz [1], Merton [1], Harrison and Pliska [1] and Aase [2].

EXAMPLE 11.5. (An optimal portfolio selection problem)
Let X_t denote the wealth of a person at time t. Suppose that the person has the choice of two different investments. The price $p_1(t)$ at time t of one of the assets is assumed to satisfy the equation

$$\frac{dp_1}{dt} = p_1(a+\alpha W_t) \tag{11.42}$$

where W_t denotes white noise and $a,\alpha>0$ are constants measuring the relative rate of change of p and the size of the noise, respectively. As we have discussed earlier we interpret (11.42) as the (Ito) stochastic differential equation

$$dp_1 = p_1 a dt + p_1 \alpha dB_t . \tag{11.43}$$

This investment is called risky, since $\alpha>0$. We assume that the price p_2 of the other asset satisfies a similar equation, but with no noise:

$$dp_2 = p_2 b dt \tag{11.44}$$

This investment is called safe. So it is natural to assume $b<a$. At each instant the person can choose how big fraction u of his wealth he will invest in the risky asset, thereby investing the fraction $1-u$ in the safe one. This gives the following stochastic differential equation for the wealth $X_t=X_t^u$:

$$\begin{aligned} dX_t &= uX_t a dt + uX_t \alpha dB_t + (1-u)X_t b dt \\ &= X_t(au+b(1-u))dt + \alpha uX_t dB_t \end{aligned} \tag{11.45}$$

Suppose that, starting with the wealth $X_t=x>0$ at time t, the person wants to maximize the expected utility of the wealth at some future time $t_0>t$. If we allow no borrowing (i.e. require $X>0$) and are given an utility function $U: [0,\infty)\to[0,\infty)$, $U(0)=0$ (usually assumed to be increasing and concave) the problem is to find a (Markov) control $u^*=u^*(t,X_t)$, $0\le u^*\le 1$, such that

$$\sup\{M^u;\ u \text{ Markov control, } 0\le u\le 1\} = M^{u^*} , \tag{11.46}$$

where $M^u = E^{t,x}[U(X_\tau^u)]$ and τ is the first exit time from the region $\{(s,z);\ s<t_0,\ z>0\}$. By putting $J^u=-M^u$ we see that this is a performance criterion of the form (11.4)-(11.5) with $F=0$ and $K=-U$. The infinitesimal generator A^v has the form (see (11.7))

$$(11.47)\qquad (A^v f)(t,x) = \frac{\partial f}{\partial t} + x(av+b(1-v))\frac{\partial f}{\partial x} + \frac{1}{2}\alpha^2 v^2 x^2 \frac{\partial^2 f}{\partial x^2}$$

The HJB equation becomes

$$(11.48)\qquad \inf_v\{(A^v H)(t,x)\} = 0\ ,\quad H(t_0,x) = -U(x)$$

or

$$(11.49)\qquad \sup_v\{(A^v V)(t,x)\} = 0\ ,\quad V(t_0,x) = U(x)$$

where

$$(11.50)\qquad V(t,x) = -H(t,x) = -\inf_u J^u = \sup_u M^u\ .$$

In other words, for each (t,x) we try to find the value $v=u(t,x)$ which maximizes the function

$$(11.51)\qquad \eta(v) = A^v V = \frac{\partial V}{\partial t} + x(b+(a-b)v)\frac{\partial V}{\partial x} + \frac{1}{2}\alpha^2 v^2 x^2 \frac{\partial^2 V}{\partial x^2}\ .$$

If $V_x = \frac{\partial V}{\partial x} > 0$ and $V_{xx} = \frac{\partial^2 V}{\partial x^2} < 0$, the solution is

$$(11.52)\qquad v = u(t,x) = -\frac{(a-b)V_x}{x\alpha^2 V_{xx}}$$

If we substitute this in the HJB equation (11.31) we get the following non-linear boundary value problem for V:

$$(11.53)\qquad V_t + bxV_x - \frac{(a-b)^2 V_x^2}{2\alpha^2 V_{xx}} = 0 \quad \text{for}\quad t<t_0\ ,\quad x>0$$

$$(11.54)\qquad V(t,x) = U(x) \qquad\qquad \text{for}\quad t=t_0 \ \text{or}\ x=0\ .$$

The problem (11.53-54) is hard to solve for general U. An important class of increasing and concave functions are the power functions

$$(11.55)\qquad U(x) = x^r \quad \text{where}\quad 0<r<1\ .$$

If we choose such a utility function U, we try to find a solution of (11.53-54) of the form

$$V(t,x) = f(t)x^r\ .$$

Substituting we obtain

$$(11.56) \qquad V(t,x) = e^{\lambda(t_0-t)} x^r ,$$

where $\lambda = br + \frac{(a-b)^2 r}{2\alpha^2(1-r)}$.

Using (11.52) we obtain the optimal control

$$(11.57) \qquad u^*(t,x) = \frac{a-b}{\alpha^2(1-r)}$$

If $\frac{a-b}{\alpha^2(1-r)} \in (0,1)$ this is the solution to the problem, in virtue of Theorem 11.2. Note that u^* is in fact constant.

Another interesting choice of the utility function is $U(x)=\log x$, called the Kelly criterion. As noted by Aase [2] (in a more general setting) we may in this case obtain the optimal control directly by evaluating $E^{t,x}[\log(X_\tau)]$ using Dynkin's formula:

$$E^{t,x}[\log(X_\tau)] =$$
$$= \log x + E^{t,x}[\int_t^\tau \{au(s,X_s)+b(1-u(s,X_s))-\tfrac{1}{2}\alpha^2u^2(s,X_s)\}ds]$$

since $A^v(\log x)=av+b(1-v)-\frac{1}{2}\alpha^2v^2$.

So it is clear that $M^u(t,x)=E^{t,x}[\log(X_\tau)]$ is maximal if we for all s,z choose $u(s,z)$ to have the value of v which maximizes

$$av + b(1-v) - \tfrac{1}{2}\alpha^2v^2$$

i.e. we choose

$$(11.58) \qquad u(s,X_s) = \frac{a-b}{\alpha^2} \qquad \text{for all } s,\omega.$$

So this is the optimal control if the Kelly criterion is used. (Similarly, this direct method also gives the optimal control when $U(x)=x^r$. Note, however, that in both these cases more work is needed to find the corresponding maximal expected utility).

<u>EXAMPLE 11.6</u>.
Finally we include an example which shows that even quite simple - and apparently innocent - stochastic control problems can lead us beyond the reach of the theory developed in this chapter:

Suppose the system is a 1-dimensional Ito integral

$$(11.59) \qquad dX_t = dX_t^u = u(t,\omega)dB_t , \qquad X_t=x>0$$

and consider the stochastic control problem

(11.60) $$H(t,x) = \sup_u E^{t,x}[K(X_\tau)] ,$$

where τ is the first exit time from $G=\{(t,x);\ t<t_1,\ x>0\}$ for $Y_s=(t+s,X_s)$ and K is a given bounded continuous function. Intuitively, we can think of the system as the state of a game which behaves like an "excited" Brownian motion, where we can control the size u of the excitation at every instant. The purpose of the control is to maximize the expected payoff $K(X_{t_1})$ of the game at a fixed future time t_1.

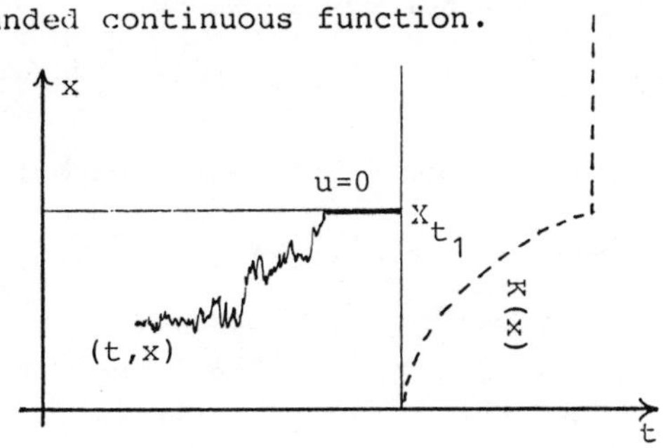

Assuming that $H\in C^2$ and that u^* exists we get by the HJB equation

(11.61) $$\sup_{v\in R}\left\{\frac{\partial H}{\partial t} + \frac{1}{2}v^2\,\frac{\partial^2 H}{\partial x^2}\right\} = 0 \quad \text{for } t<t_1 , \quad H(t_1,x)=K(x) .$$

From this we see that we necessarily have

(11.62) $$\frac{\partial^2 H}{\partial x^2} \le 0 , \quad v^*\,\frac{\partial^2 H}{\partial x^2} = 0 \quad \text{and} \quad \frac{\partial H}{\partial t} = 0 \quad \text{for } t<t_1 ,$$

where v^* is the value of $v\in\mathbb{R}$ which gives the supremum in (11.61). But if $\frac{\partial H}{\partial t} = 0$, then $H(t,x)=H(t_1,x)=K(x)$. However, this cannot possibly be the solution in general, because we have not assumed that $\frac{\partial^2 K}{\partial x^2} \le 0$ - in fact, K was not even assumed to be differentiable.

What went wrong? Since the conclusion of the HJB equation was wrong, the assumptions cannot hold. So either H is not C^2 or u^* does not exist (as a Lipschitz continuous Markov control with at most linear growth), or both.

To simplify the problem assume that

$$K(x) = \begin{cases} x^2 ; & 0\le x\le 1 \\ 1 ; & x>1 \end{cases}$$

Then considering the figure above and using some intuition we see that it is optimal to excite as much as possible if X_t is in the strip $0<x<1$ to avoid exiting from G in the interval $\{t_1\}\times(0,1)$. Using that X_t is just a time change of Brownian motion (see Chapter VIII) we conclude that this optimal control leads to a process X^*

which jumps immediately to the value 1 with probability x and to the value 0 with probability $1-x$, if the starting point is $x\in(0,1)$. If the starting point is $x\in[1,\infty)$ we simply choose our control to be zero. In other words, we put

$$(11.63)\qquad u^*(t,x) = \begin{cases} \infty & \text{if } x\in(0,1) \\ 0 & \text{if } x\in[1,\infty) \end{cases}$$

with corresponding expected payoff

$$(11.64)\qquad h^*(t,x) = E^{t,x}[K(X^*_{t^*_1})] = \begin{cases} x & \text{if } 0\le x\le 1 \\ 1 & \text{if } x>1 \end{cases}$$

Thus we see that our candidate u^* for optimal control is not continuous (not even finite!) and the corresponding optimal process X^*_t is not an Ito diffusion (it is not even continuous). So to handle this case mathematically it is necessary to enlarge the family if admissible controls (and the family of corresponding processes). For example, one can prove an extended version of Theorem 11.2 which allows us to conclude that our choice of u^* above does indeed give at least as good performance as any other Markov control u and that h^* given by (11.64) does coincide with the maximal expected payoff H defined by (11.60).

This last example illustrates the importance of the question of existence in general, both of the optimal control u^* and of the corresponding solution X_t of the stochastic differential equation (11.1). We briefly outline some results in this direction:

With certain conditions on $b,\sigma,F,\partial G$ and assuming that the set of control values is compact, one can show, using general results from non-linear partial differential equations, that a smooth function h exists such that

$$\inf_v\{F^v(y)+(A^v h)(y)\} = 0 \qquad \text{for } y\in G$$

and

$$h(y) = K(y) \qquad \text{for } y\in\partial G\ .$$

Then by a measurable selection theorem one can find a (measurable) function $u^*(y)$ such that

$$(11.65)\qquad F^{u^*}(y) + (A^{u^*}h)(y) = 0\ ,$$

for a.a. $y\in G$ wrt. Lebesgue measure in $\mathbb{R}^k$. Even if u^* is only known to be measurable, one can show that the corresponding solution

$X_t = X_t^{u^*}$ of (11.1) exists (see Stroock and Varadhan [1] for general results in this direction). Then by inspecting the proof of Theorem 11.2 one can see that it suffices to have (11.65) satisfied outside a subset of G with Green measure 0 (see Definition 9.19). Under suitable conditions on b and σ one can in fact show that the Green measure is absolutely continuous wrt. Lebesgue measure. Thus by (11.65) (and a strengthened Theorem 11.2) u^* is an optimal control. We refer the reader to Fleming and Rishel [1], Dynkin and Yushkevich [1] and Krylov [1] for details and further studies.

Appendix A: Normal Random Variables

Here we recall some basic facts which are used in the text.

DEFINITION A.1. Let $(\Omega, \mathcal{F}, P)$ be a probability space. A random variable $X : \Omega \to \mathbb{R}$ is normal if the distribution of X has a density of the form

(A.1) $$p_X(x) = \frac{1}{\sigma\sqrt{2\pi}} \cdot \exp\left(-\frac{(x-m)^2}{2\sigma^2}\right),$$

where $\sigma > 0$ and m are constants. In other words,

$$P[X \in G] = \int_G p_X(x)dx, \quad \text{for all Borel sets } G \subset \mathbb{R}$$

If this is the case, then

(A.2) $$E(X) = \int_\Omega X dP = \int_R x p_X(x)dx = m$$

and

(A.3) $$\text{var}[X] = E[(X-m)^2] = \int_R (x-m)^2 p_X(x)dx = \sigma^2$$

More generally, a random variable $X : \Omega \to \mathbb{R}^n$ is called (multi) normal $N(m,C)$ if the distribution of X has a density of the form

(A.4) $$p_X(x_1,\ldots,x_n) = \frac{\sqrt{|A|}}{(2\pi)^{n/2}} \cdot \exp\left(-\frac{1}{2} \cdot \sum_{j,k} (x_j-m_j)a_{jk}(x_k-m_k)\right)$$

where $m = (m_1,\ldots,m_n) \in \mathbb{R}^n$ and $C^{-1} = A = [a_{jk}] \in \mathbb{R}^{n\times n}$ is a positive definite matrix.

If this is the case then

(A.5) $$E[X] = m$$

and

(A.6) $$A^{-1} = C = [c_{jk}] \text{ is the covariance matrix of } X, \text{ i.e.}$$

$$c_{jk} = E[(X_j-m_j)(X_k-m_k)]$$

DEFINITION A.2 The characteristic function of a random variable $X : \Omega \to \mathbb{R}^n$ is the function $\phi_X : \mathbb{R}^n \to \mathbb{C}$ (where $\mathbb{C}$ denotes the complex numbers) defined by

(A.7) $$\phi_X(u_1,\ldots,u_n) = E[\exp(i(u_1X_1+\ldots+u_nX_n))] = \int_{\mathbb{R}^n} e^{i\langle u,x\rangle} \cdot P[X\in dx],$$

where $\langle u,x\rangle = u_1x_1+\ldots+u_nx_n$ (and $i \in \mathbb{C}$ is the imaginary unit). In other words, ϕ_X is the Fourier transform of X (or, more precisely, of the measure $P[X\in dx]$). Therefore we have

THEOREM A.3. The characteristic function of X determines the distribution of X uniquely.

It is not hard to verify the following:

THEOREM A.4. If $X : \Omega \to \mathbb{R}^n$ is normal $N(m,C)$, then

$$\phi_X(u_1,..,u_n) = \exp(-\frac{1}{2} \sum_{j,k} u_j c_{jk} u_k + i \sum_j u_j m_j) \tag{A.8}$$

Theorem A.4 is often used as a basis for an extended concept of a normal random variable: We define $X:\Omega \to \mathbb{R}^n$ to be normal (in the extended sense) if ϕ_X satisfies (A.8) for some symmetric non-negative definite matrix $C = [c_{jk}] \in \mathbb{R}^{n \times n}$ and some $m \in \mathbb{R}^n$. So by this definition it is not required that C be invertible. From now on we will use this extended definition of normality.
In the text we often use the following result:

THEOREM A.5. Let $X_j : \Omega \to \mathbb{R}$ be random variables; $1 \leq j \leq n$. Then

$$X = (X_1,...,X_n) \text{ is normal}$$

if and only if

$$Y = \lambda_1 X_1 + \dots + \lambda_n X_n \text{ is normal for all } \lambda_1,..,\lambda_n \in \mathbb{R}.$$

Proof. If X is normal, then

$$E[\exp(iu(\lambda_1 X_1 + .. + \lambda_n X_n))] = \exp(-\frac{1}{2} \sum_{j,k} u\lambda_j c_{jk} u\lambda_k + i\sum_j u \lambda_j m_j)$$

$$= \exp(-\frac{1}{2} \cdot u^2 \sum_{j,k} \lambda_j c_{jk} \lambda_k + i \cdot u \cdot \sum_j \lambda_j m_j),$$

so Y is normal with $E[Y] = \sum \lambda_j m_j$, $\mathrm{var}[Y] = \sum \lambda_j c_{jk} \lambda_k$.

Conversely, if $Y = \lambda_1 X_1 + .. + \lambda_n X_n$ is normal with $E[Y] = m$ and $\mathrm{var}[Y] = \sigma^2$, then

$$E[\exp(iu(\lambda_1 X_1 + ... + \lambda_n X_n))] = \exp(-\frac{1}{2} u^2 \sigma^2 + i u m),$$

where $m = \sum_j \lambda_j E[X_j]$, $\sigma^2 = E[(\sum_j \lambda_j X_j - \sum_j \lambda_j E[X_j])^2]$

$$= E[(\sum_j \lambda_j (X_j - m_j))^2] = \sum_{j,k} \lambda_j \lambda_k E[(X_j - m_j)(X_k - m_k)],$$

where $m_j = E[X_j]$. Hence X is normal.

THEOREM A.6. Let $Y_0, Y_1, \ldots, Y_n$ be real, random variables on Ω. Assume that $X = (Y_0, Y_1, \ldots, Y_n)$ is normal and that Y_0 and Y_j are uncorrelated for each $j \geq 1$, i.e.

$$E[(Y_0 - E[Y_0])(Y_j - E[Y_j])] = 0 \; ; \; 1 \leq j \leq n.$$

Then Y_0 is independent of $\{Y_1, \ldots, Y_n\}$.

Proof. We have to prove that

$$(A.9) \quad P[Y_0 \in G_0, Y_1 \in G_1, \ldots, Y_n \in G_n] = P[Y_0 \in G_0] \cdot P[Y_1 \in G_1, \ldots, Y_n \in G_n],$$

for all Borel sets $G_0, G_1, \ldots, G_n \subset \mathbb{R}$.
We know that in the first line (and the first column) of the covariance matrix $c_{jk} = E[(Y_j - E[Y_j])(Y_k - E[Y_k])]$ only the first entry $c_{00} = \text{var}[Y_0]$, is non-zero. Therefore the density of X satisfies

$$p_X(x_0, \ldots, x_n) = p_{Y_0}(x_0) \cdot p_{Y_1, \ldots, Y_n}(x_1, \ldots, x_n),$$

and this is equivalent to (A.9).

Finally we establish the following:

THEOREM A.7. Suppose $X_k : \Omega \to \mathbb{R}^n$ is normal for all k and that $X_k \to X$ in $L^2(\Omega)$, i.e.

$$E[|X_k - X|^2] \to 0 \quad \text{as} \quad k \to \infty$$

Then X is normal.

Proof. Since $|e^{i\langle u,x\rangle} - e^{i\langle u,y\rangle}| \leq |u| \cdot |x-y|$, we have

$$E[\{\exp(i\langle u, X_k\rangle) - \exp(i\langle u, X\rangle)\}^2] \leq |u|^2 \cdot E[|X_k - X|^2] \to 0 \quad \text{as} \quad k \to \infty.$$

Therefore

$$E[\exp(i\langle u, X_k\rangle)] \to E[\exp(i\langle u, X\rangle)] \quad \text{as} \quad k \to \infty.$$

So X is normal, with mean $E[X] = \lim E[X_k]$ and covariance matrix $C = \lim C_k$, where C_k is the covariance matrix of X_k.

Appendix B: Conditional Expectations

Let $(\Omega, \mathcal{F}, P)$ be a probability space and let $X : \Omega \to \mathbb{R}^n$ be a random variable such that $E[|X|] < \infty$. If $\mathcal{H} \subset \mathcal{F}$ is a σ-algebra, then <u>the conditional expectation of X given $\mathcal{H}$</u>, denoted by $E[X|\mathcal{H}]$, is defined as follows:

<u>DEFINITION B.1</u>. $E[X|\mathcal{H}]$ is the (a.s. unique) function from Ω to $\mathbb{R}^n$ satisfying:

(1) $\quad E[X|\mathcal{H}]$ is $\mathcal{H}$-measurable

(2) $\quad \int_A E[X|\mathcal{H}]dP = \int_A X\, dP$, for all $A \in \mathcal{H}$.

The existence and uniqueness of $E[X|\mathcal{H}]$ comes from the Radon-Nikodym theorem:

Let μ be the measure on $\mathcal{H}$ defined by

$$\mu(A) = \int_A X\, dP \ ; \ A \in \mathcal{H}.$$

Then μ is absolutely continuous wrt. $P|\mathcal{H}$, so there exists a $P|\mathcal{H}$-unique $\mathcal{H}$-measurable function F on Ω such that

$$\mu(A) = \int_A F\, dP \quad \text{for all } A \in \mathcal{H}.$$

Thus $E[X|\mathcal{H}] = F$ does the job and this function is unique a.s. wrt. the measure $P|\mathcal{H}$.

We list some of the basic properties of the conditional expectation:

<u>THEOREM B.2</u>. Suppose $Y : \Omega \to \mathbb{R}^n$ is another random variable with $E[|Y|] < \infty$ and let $a, b \in \mathbb{R}$. Then

a) $\quad E[aX + bY|\mathcal{H}] = aE[X|\mathcal{H}] + bE[Y|\mathcal{H}]$

b) $\quad E[E[X|\mathcal{H}]] = E[X]$

c) $\quad E[X|\mathcal{H}] = X$ if X is $\mathcal{H}$-measurable

d) $\quad E[X|\mathcal{H}] = E[X]$ if X is independent of $\mathcal{H}$

e) $\quad E[Y\cdot X|\mathcal{H}] = Y\cdot E[X|\ ]$ if Y is $\mathcal{H}$-measurable, where $\cdot$ denotes the usual inner product in $\mathbb{R}^n$.

<u>Proof</u>. d): If X is independent of $\mathcal{H}$ we have for $A \in \mathcal{H}$

$$\int_A X dP = \int_\Omega X\cdot \chi_A \cdot dP = \int_\Omega X\cdot dP\cdot \int_\Omega \chi_A dP = E[X]\cdot P(A),$$

so the constant $E[X]$ satisfies (1) and (2).

e): We first establish the result in the case when $Y = \chi_B$ (where χ denotes the indicator function), for some $B \in \mathcal{H}$.

Then for all $A \in \mathcal{H}$

$$\int_A Y \cdot E[X|\mathcal{H}]dP = \int_{A\cap B} E[X|\mathcal{H}]dP = \int_{A\cap B} XdP = \int_A YX\, dP,$$

so $Y \cdot E[X|\mathcal{H}]$ satisfies both (1) and (2). Similarly we obtain that the result is true if Y is a simple function

$$Y = \sum_{j=1}^{m} c_j \chi_{B_j}, \quad \text{where} \quad B_j \in \mathcal{H}.$$

The result in the general case then follows by approximating Y by such simple functions.

THEOREM B.3. Let $\mathcal{G}, \mathcal{H}$ be σ-algebras such that $\mathcal{G} \subset \mathcal{H}$. Then

$$E[X|\mathcal{G}] = E[E[X|\mathcal{H}]|\mathcal{G}].$$

Proof. If $A \in \mathcal{G}$ then $A \in \mathcal{H}$ and therefore

$$\int_A E[X|\mathcal{H}]dP = \int_A X\, dP.$$

Hence $E[E[X|\mathcal{H}]|\mathcal{G}] = E[X|\mathcal{G}]$ by uniqueness.

Appendix C: Uniform Integrability and Martingale Convergence

We give a brief summary of the definitions and results which are the background for the applications in this book. For proofs and more information we refer to Doob [1], Liptser and Sheryayev [1], Meyer [1] or Williams [1].

DEFINITION C.1. Let $(\Omega, \mathcal{F}, P)$ be a probability space. A family $\{f_j\}_{j\in J}$ of real, measurable functions f_j on Ω is called uniformly integrable if

$$\lim_{n\to\infty} \left(\sup_{j\in J} \left\{ \int_{\{|f_j|>n\}} |f_j| dP \right\}\right) = 0$$

One of the most useful tests for uniform integrability is obtained by using the following concept:

DEFINITION C.2. A function $\psi : [0,\infty) \to [0,\infty)$ is called a u.i. (uniform integrability) test function if ψ is increasing, convex (i.e. $\psi(\lambda x+(1-\lambda)y) \leq \lambda\psi(s)+(1-\lambda)\psi(y)$ for all $x,y \in [0,\infty)$, $\lambda \in [0,1]$) and

$$\lim_{x\to\infty} \frac{\psi(x)}{x} = \infty.$$

So for example $\psi(x) = x^p$ is a u.i. test function if $p > 1$, but not if $p = 1$.

The justification for the name in Definition C.2 is the following:

THEOREM C.3. The family $\{f_j\}_{j\in J}$ is uniformly integrable if and only if there is a u.i. test function ψ such that

$$\sup_{j\in J} \left\{\int \psi(|f_j|)dP\right\} < \infty$$

One major reason for the usefulness of uniform integrability is the following result, which may be regarded as the ultimate generalization of the various convergence theorems in integration theory:

THEOREM C.4. Suppose $\{f_k\}_{k=1}^{\infty}$ is a sequence of real measurable functions on Ω such that

$$\lim_{k\to\infty} f_k(\omega) = f(\omega) \quad \text{for a.a. } \omega.$$

Then the following are equivalent:

1) $\{f_k\}$ is uniformly integrable

2) $f \in L^1(P)$ and $f_k \to f$ in $L^1(P)$,

   i.e. $\int |f_k - f| dP \to 0$ as $k \to \infty$.

An important application of uniform integrability is within the convergence theorems for martingales:

Let $(\Omega, \mathcal{N}, P)$ be a probability space and let $\{\mathcal{N}_t\}_{t \geq 0}$ be an increasing family of σ-algebras, $\mathcal{N}_t \subset \mathcal{N}$ for all t. A stochastic process $N_t: \Omega \to \mathbb{R}$ is called a <u>supermartingale</u> (wrt. $\{\mathcal{N}_t\}$) if N_t is $\mathcal{N}_t$-adapted, $E[|N_t|] < \infty$ for all t and

(C.1) $\quad N_t \geq E[N_s | \mathcal{N}_t] \quad$ for all $s > t$

(Similarly, if (C.1) holds with the inequality reversed for all $s>t$, then N_t is called a <u>submartingale</u>. And if (C.1) holds with equality then N_t is called a <u>martingale</u>).

As in customary we will assume that each $\mathcal{N}_t$ contains all the null sets of $\mathcal{N}$, that $t \to N_t(\omega)$ is right continuous for a.a.ω and that $\{\mathcal{N}_t\}$ is right continous, in the sense that $\mathcal{N}_t = \bigcap_{s>t} \mathcal{N}_s$ for all $t \geq 0$.

<u>THEOREM C.5</u>. (Doob's martingale convergence theorem I)
Let N_t be a right continuous supermartingale with the property that

$$\sup_{t>0} E[N_t^-] < \infty,$$

where $N_t^- = \max(-N_t, 0)$.
Then the pointwise limit

$$N(\omega) = \lim_{t\to\infty} N_t(\omega)$$

exists for a.a.ω and $E[N^-] < \infty$.

Note, however, that the convergence need not be in $L^1(P)$. In order to obtain this we need uniform integrability:

<u>THEOREM C.6</u>. (Doob's martingale convergence theorem II).
Let N_t be a right-continuous supermartingale. Then the following are equivalent:

1) $\{N_t\}_{t \geq 0}$ is uniformly integrable

2) There exists $N \in L^1(P)$ such that $N \to N$ a.e. (P) and $N_t \to N$ in $L^1(P)$, i.e. $\int |N_t - N| dP \to 0$ as $t \to \infty$.

Combining Theorems C.6 and C.3 (with $\psi(x) = x^p$) we get

COROLLARY C.7. Let M_t be a continuous martingale such that

$$\sup_{t>0} E[|M_t|^p] < \infty \quad \text{for some } p>1.$$

Then there exists $M \in L^1(P)$ such that $M_t \to M$ a.e. (P) and

$$\int |M_t - M| dP \to 0 \quad \text{as } t \to \infty.$$

Finally, we mention that similar results can be obtained for the analogous discrete time super/sub-martingales $\{N_k, \mathcal{N}_k\}$, $k=1,2,\ldots$. Of course, no continuity assumptions are needed in this case. For example, we have the following result, which is used in Chapter IX:

COROLLARY C.8. Let M_k; $k=1,2,\ldots$ be a discrete time martingale and assume that

$$\sup_{k} E[|M_k|^p] < \infty \quad \text{for some } p>1.$$

Then there exists $M \in L^1(P)$ such that $M_k \to M$ a.e. (P) and

$$\int |M_k - M| dP \to 0 \quad \text{as } k \to \infty.$$

Bibliography

K.K. AASE:

[1] Stochastic continuous-time model reference adaptive systems with decreasing gain. Advances in Appl. Prop. 14(1982), 763-788.

[2] Optimum portfolio diversification in a general continuous time model. Stochastic Processes and their Applications 18 (1984), 81-98.

L. ARNOLD:

[1] Stochastische Differentialgleichungen. Theorie und Anwendung. R. Oldenbourgh Verlag 1973.

J.A. BATHER:

[1] Optimal stopping problems for brownian motion. Advances in Appl. Prob. 2 (1970), 259-286.

A. BENSOUSSAN AND J.L. LIONS:

[1] Applications des inéquations variationelles en controle stochastique. Dunod 1978.

A. BERNARD, E.A. CAMPBELL AND A.M. DAVIE:

[1] Brownian motion and generalized analytic and inner functions. Ann. Inst. Fourier 29 (1979), 207-228.

L. BERS, F. JOHN AND M. SCHECHTER:

[1] Partial Differential Equations. Interscience 1964.

R.M. BLUMENTHAL AND R.K. GETOOR:

[1] Markov Processes and Potential Theory. Academic Press 1968.

L. BREIMAN:

[1] Probability. Addison-Wesley 1968.

B.M. BROWN AND J.I. HEWITT:

[1] Asymptotic likelihood theory for diffusion processes. J. Appl. Prob. 12 (1975), 228-238.

R.S. BUCY AND P.D. JOSEPH:

[1] Filtering for Stochastic Processes with Applications to Guidance. Interscience Publ. 1968.

Y.S. CHOW, H. ROBBINS, D. SIEGMUND:

[1] Great Expectations: The Theory of Optimal Stopping. Houghto Miffin Co., New York 1971.

K.L. CHUNG:

[1] Lectures from Markov Processes to Brownian Motion. Springer-Verlag 1982.

K.L. CHUNG AND R. WILLIAMS:

[1] Introduction to Stochastic Integration Birkhäuser 1983.

L. CSINK AND B. ØKSENDAL:

[1] Stochastic harmonic morphisms: Functions mapping the paths of one diffusion into the paths of another. Ann. Inst. Fourier 33 (1983), 219-240.

M.H.A. DAVIS:

[1] Linear Estimation and Stochastic Control. Chapman & Hall 1977.

[2] Lectures on Stochastic Control and Nonlinear Filtering Tata Institute of Fundamental Research 75, Springer-Verlag 1984.

J.L. DOOB:

[1] Classical Potential Theory and Its Probabilistic Counterpart Springer-Verlag 1984.

R. DURRETT:

[1] Brownian Motion and Martingales in Analysis. Wadsworth Inc. 1984.

E.B. DYNKIN:

[1] The optimum choice of the instant for stopping a Markov process. Soviet Mathematics 4 (1963), 627-629.

[2] Markov Processes, vol. I. Springer-Verlag 1965.

[3] Markov Processes, vol. II. Springer-Verlag 1965.

E.B. DYNKIN AND A.A. YUSHKEVICH:

[1] Controlled Markov Processes. Springer-Verlag 1979.

R.J. ELLIOTT:

[1] Stochastic Calculus and Applications. Springer-Verlag 1982.

K.D. ELWORTHY:

[1] Stochastic Differential Equations on manifolds. Cambridge University Press 1982.

W.H. FLEMING AND R.W. RISHEL:

[1] Deterministic and Stochastic Optimal Control. Springer-Verlag 1975.

A. FRIEDMAN:

[1] Stochastic Differential Equations and Applications, vol. I. Academic Press 1975.

[2] Stochastic Differential Equations and Applications, vol II. Academic Press 1976.

M. FUKUSHIMA:

[1] Dirichlet Forms and Markov Processes. North Holland/Kodansha 1980.

I.I. GIHMAN AND A.V. SKOROHOD:

[1] Stochastic Differential Equations. Springer-Verlag 1974.

[2] The Theory of Stochastic Processes, vol. I. Springer-Verlag 1974.

[3] The Theory of Stochastic Processes, vol. II. Springer-Verlag 1975.

[4] The Theory of Stochastic Processes, vol. III. Springer-Verlag 1979.

[5] Controlled Stochastic Processes. Springer-Verlag 1979.

J.M. HARRISON AND S.R. PLISKA:

[1] Martingales and stochastic integrals in the theory of continuous trading. Stochastic Processes and their Applications 11 (1981), 215-260.

T. HIDA:

[1] Brownian Motion. Springer-Verlag 1980.

K. HOFFMAN:

[1] Banach Spaces of Analytic Functions. Prentice Hall 1962.

N. IKEDA AND S. WATANABE:

[1] Stochastic Differential Equations and Diffusion Processes. North Holland/Kodansha 1981.

K. ITO AND H.P. McKEAN:

[1] Diffusion Processes and Their Sample Paths. Springer-Verlag 1965.

A.H. JASWINSKI:

[1] Stochastic Processes and Filtering Theory. Academic Press 1970.

G. KALLIANPUR:

[1] Stochastic Filtering Theory. Springer-Verlag 1980.

S. KARLIN AND H. TAYLOR:

[1] A First Course in Stochastic Processes (2. edition). Academic Press 1975.

[2] A Second Course in Stochastic Processes. Academic Press 1981.

F.B. KNIGHT:

[1] Essentials of Brownian Motion. American Math. Soc. 1981.

V. KRISHNAN:

[1] Nonlinear Filtering and Smoothing: An Introduction to Martingales, Stochastic Integrals and Estimation. J. Wiley 1984.

N.V. KRYLOV:

[1] Controlled Diffusion Processes. Springer-Verlag 1980.

N.V. KRYLOV AND A.K. ZVONKIN:

[1] On strong solutions of stochastic differential equations. Sel.Math.Sov. I (1981), 19-61.

H.J. KUSHNER:

[1] Stochastic Stability and Control. Academic Press 1967.

J. LAMPERTI:

[1] Stochastic Processes. Springer-Verlag 1977.

R.S. LIPSTER AND A.N. SHIRYAYEV:

[1] Statistics of Random Processes, vol. I. Springer-Verlag 1977.

[2] Statistics of Random Processes, vol. II. Springer-Verlag 1978.

R. McDONALD AND D. SIEGEL:

[1] The value of waiting to invest. Quarterly J. of Economics 101 (1986), 707-727.

T.P. McGARTY:

[1] Stochastic Systems and State Estimation. J. Wiley & Sons 1974.

H.P. McKEAN:

[1] A free boundary problem for the heat equation arising from a problem of mathematical economics. Industrial managem. review 6 (1965), 32-39.

[2] Stochastic Integrals. Academic Press 1969.

A.G. MALLIARIS:

[1] Ito's calculus in financial decision making. SIAM Review 25 (1983), 481-496.

A.G. MALLIARIS AND W.A. BROCK

[1] Stochastic Methods in Economics and Finance. North-Holland 1982.

H.M. MARKOWITZ:

[1] Portolio Selection. Efficient Diversification of Investments. Yale University Press 1976.

R.C. MERTON:

[1] Optimum consumption and portfolio rules in a continuous-time model. Journal of Economic Theory 3 (1971), 373-413.

M. METIVIER AND J. PELLAUMAIL:

[1] Stochastic Integration. Academic Press 1980.

P.A. MEYER:

[1] Probability and Potentials. Blaisdell 1966.

[2] Un cours sur les intégrales stochastiques. Sem. de Prob. X, Springer LNM 511 (1976), 245-400.

B. ØKSENDAL:

[1] Finely harmonic morphisms, Brownian path preserving functions and conformal martingales. Inventiones math. 75 (1984), 179-187.

[2] Stochastic processes, infinitesimal generators and function theory. In "Operators and Function Theory", NATO Advanced Study Institute Proceedings, D. Reidel 1985.

B. ØKSENDAL:

[3] When is a stochastic integral a time change of a diffusion? Preprint, University of Oslo 1988 (to appear).

T.E. OLSEN AND G. STENSLAND:

[1] A note on the value of waiting to invest. Manuscript 1987, CMI, N-5036 Fantoft, Norway.

S. PORT AND C. STONE:

[1] Brownian Motion and Classical Potential Theory. Academic Press 1979.

F.P. RAMSEY:

[1] A mathematical theory of saving. Economic J. 38 (1928), 543-549.

M. RAO:

[1] Brownian Motion and Classical Potential Theory. Aarhus Univ. Lecture Notes in Math. 47, 1977.

L.C.G. ROGERS AND D. WILLIAMS:

[1] Diffusions, Markov Processes, and Martingales, vol. 2. J. Wiley & Sons 1987.

YU.A. ROZANOV:

[1] Markov Random Fields. Springer-Verlag 1982.

P.A. SAMUELSON:

[1] Rational theory of warrant pricing. Industrial managem. review 6 (1965), 13-32.

A.N. SHIRYAYEV:

[1] Optimal Stopping Rules. Springer-Verlag 1978.

B. SIMON:

[1] Functional Integration and Quantum Physics. Academic Press 1979.

J.L. SNELL:

[1] Applications of martingale system theorems. Trans. Amer. Math. Soc. 73 (1952), 293-312.

R.L. STRATONOVICH:

[1] A new representation for stochastic integrals and equations. J. Siam Control 4 (1966), 362-371.

D.W. STROOCK:

[1] On the growth of stochastic integrals Z. Wahr. verw. Geb. 18 (1971), 340-344.

[2] Topics in Stochastic Differential Equations. Tata Institute of Fundamental Research/Springer-Verlag 1981.

D.W. STROOCK AND S.R.S. VARADHAN:

[1] Multidimensional Diffusion Processes. Springer-Verlag 1979.

H.J. SUSSMANN:

[1] On the gap between deterministic and stochastic ordinary differential equations. The Annals of Prob. 6 (1978), 19-41.

A. TARASKIN:

[1] On the asymptotic normality of vektorvalued stochastic integrals and estimates of drift parameters of a multidimensional diffusion process. Theory Prob. Math. Statist. 2 (1974), 209-224.

THE OPEN UNIVERSITY:

[1] Mathematical models and methods, unit 11. The Open University Press 1981.

F. TOPSØE:

[1] An information theoretical game in connection with the maximum entropy principle (Danish). Nordisk Matematisk Tidsskrift 25/26 (1978), 157-172.

M. TURELLI:

[1] Random environments and stochastic calculus. Theor. Pop. Biology 12 (1977), 140-178.

P. VAN MOERBEKE:

[1] An optimal stopping problem with linear reward. Acta Mathematica 132 (1974), 111-151.

D. WILLIAMS:

[1] Diffusions, Markov Processes and Martingales, J. Wiley & Sons 1979.

D. WILLIAMS (editor):

[1] Stochastic Integrals. Springer LNM. 851, Springer-Verlag 1981.

E. WONG:

[1] Stochastic Processes in Information and Dynamical Systems. McGraw Hill 1971.

E. WONG AND M. ZAKAI:

[1] Riemann-Stieltjes approximations of stochastic integrals. Z. Wahr. verw.Geb. 12 (1969), 87-97.

List of Frequently used Notation and Symbols

$\mathbb{R}^n$	n-dimensional Euclidean space
$\mathbb{R}^+$	the non-negative real numbers
$\mathbb{Z}$	the integers
$\mathbb{Z}^+ = \mathbb{N}$	the natural numbers
$\mathbb{C}$	the complex plane
$\mathbb{R}^{n\times m}$	the $n \times m$ matrices (real entries)
$\mathbb{R}^n \simeq \mathbb{R}^{n\times 1}$	, i.e. vectors in $\mathbb{R}^n$ are regarded as $n\times 1$- matrices.
$\lvert x \rvert$	$(\sum_{i=1}^n x_i^2)^{1/2}$ if $x \in \mathbb{R}^n$.
$C(U,V)$	the continuous functions from U into V
$C(U)$	the same as $C(U, \mathbb{R})$
$C_0(U)$	the functions in $C(U)$ with compact support
$C^k = C^k(U)$	the functions in $C(U, \mathbb{R})$ with continuous derivatives up to order k.
$C^{k+\alpha}$	the functions in C^k whose k'th derivatives are Lipschitz continuous with exponent α
$C_b(U)$	the bounded continuous functions on U
$f\vert K$	the restriction of the function f to the set K
A	the generator of an Ito diffusion
$\mathscr{A}$	the characteristic operator of an Ito diffusion
B_t	Brownian motion
$\mathscr{D}_A$	domain of definition of the operator A
Δ	the Laplace operator: $\Delta f = \sum_i \frac{\partial^2 f}{\partial x_i^2}$.
L	a semielliptic second order partial differential operator of the form $L = \sum_i b_i \frac{\partial}{\partial x_i} + \sum_{i,j} a_{ij} \frac{\partial^2}{\partial x_i \partial x_j}$.
R_α	the resolvent operator
iff	if and only if
a.a., a.e., a.s.	almost all, almost everywhere, almost surely
wrt.	with respect to
s.t.	such that

$E[Y] = E^{\mu}[Y] = \int Y d\mu$	expectation of the random variable Y wrt. the measure μ.
$E[Y\|\mathcal{N}]$	conditional expectation of Y wrt. $\mathcal{N}$
$\mathcal{B}$	the Borel σ-algebra
$\mathcal{F}_t$	σ-algebra generated by $\{B_s;\ s\leq t\}$
$\mathcal{M}_t$	σ-algebra generated by $\{X_s;\ s\leq t\}$ (X_t an Ito diffusion)
∂G	the boundary of the set G
$\bar{G}$	the closure of the set G
$G \subset\subset H$	$\bar{G}$ is compact and $\bar{G} \subset H$.
τ_G	the first exit time from the set G of a process X_t: $\tau_G = \inf\{t>0;\ X_t \notin G\}$.
HJB	the Hamilton-Jacobi-Bellman equation
I_n	the $n \times n$ identity matrix
χ_G	the indicator function of the set G
P^x	the probability law of B_t starting at x
Q^x	the probability law of X_t starting at x
$R^{(t,x)}$	the probability law of Y_s starting at (t,x) (Chapter X)
$s \wedge t$	the minimum of s and t
$s \vee t$	the maximum of s and t
σ^T	the transposed of the matrix σ
δ_x	the unit point mass at x
θ_t	the shift operator: $\theta_t(f(X_s)) = f(X_{t+s})$ (Chapter VII).

Index

M. Emery, University of Strasbourg, France

Stochastic Calculus in Manifolds

1989. X, 153 pp. 3 figs. Softcover, in preparation.
ISBN 3-540-51664-6

Contents: Real Semimartingales and Stochastic Integrals. - Some Vocabulary from Differential Geometry. - Manifold-valued Semimartingales and their Quadratic Variation. - Connections and Martingales. - Riemannian Manifolds and Brownian Motions. - Second Order Vectors and Forms. - Stratonovich and Itô Integrals of First Order Forms. - Parallel Transport and Moving Frame. - Appendix: A Short Presentation of Stochastic Calculus.

Springer-Verlag Berlin
Heidelberg New York London
Paris Tokyo Hong Kong